DISCARD

Problem Books in Mathematics

Edited by P. R. Halmos

Problem Books in Mathematics

Series Editor: P.R. Halmos

(continued after index)

Nathan Keyfitz
John A. Beekman

Demography
Through Problems

With 22 Illustrations

Springer-Verlag
New York Berlin Heidelberg Tokyo

Nathan Keyfitz
International Institute for
   Applied Systems Analysis
Laxembourg, Austria

John A. Beekman
Department of Mathematical Sciences
Ball State University
Muncie, IN 47306
U.S.A.

Editor
Paul R. Halmos
Department of Mathematics
Santa Clara University
Santa Clara, CA 95053
U.S.A.

Mathematical Subject Classifications: 00A07, 60J70, 60J80, 62P10, 92A15

Library of Congress Cataloging in Publication Data
Keyfitz, Nathan, 1913–
   Demography through problems.
   (Problem books in mathematics)
   Bibliography: p.
   Includes index.
   1. Demography—Problems, exercises, etc.   I. Beekman,
John A.   II. Title.   III. Series.
HB850.3.K49   1984   304.6′076   83-6775

Typeset by Composition House Ltd., Salisbury, England.
Printed and bound by R. R. Donnelley & Sons, Harrisonburg, Virginia.
Printed in the United States of America.

9 8 7 6 5 4 3 2

ISBN 0-387-90836-6 Springer-Verlag New York Berlin Heidelberg Tokyo
ISBN 3-540-90836-6 Springer-Verlag Berlin Heidelberg New York Tokyo

Preface

The book that follows is an experiment in the teaching of population theory and analysis. A sequence of problems where each is a self-contained puzzle, and the successful solution of each which puts the student in a position to tackle the next, is a means of securing the active participation of the learner and so the mastery of a technical subject. How far our questions are the exciting puzzles at which we aimed, and how far the sequence constitutes a rounded course in demography, must be left to the user to judge. One test of a good problem is whether a solution, that may take hours of cogitation, is immediately recognizable once it comes to mind.

While algebraic manipulation is required throughout, we have tried to emphasize problems in which there is some substantive point—a conclusion regarding population that can be put into words. Our title, *Demography Through Problems*, reflects our intention of leading the reader who will actively commit him- or herself through a sequence that will not only teach definitions—in itself a trivial matter—but sharpen intuition on the way that populations behave.

We experienced tension between the aim of making the problems demographically relevant, on the one hand, and expressing them in succinct and self-contained formulations, on the other. In demographic practice the complexities of data and their errors, not to mention the sheer volume of material, do not readily produce incisive formulations. A book of this kind is bound to show compromises between incisiveness and the detailed minutiae of real-life data.

Actuaries as well as demographers must utilize national, state, and regional mortality statistics, cause of death statistics, and life tables, and some of the problems are designed to provide experience with the wealth of data from the U.S. Bureau of the Census and other official sources. Students

can pick a Standard Metropolitan Statistical Area (SMSA) and compute its
crude death rate, central death rates by age group and sex, and death rates
for various causes. As one attempt to explain the revealed differences,
students can compute adjusted death rates and age–sex-adjusted death rates
for their SMSA's, where the standard population figures were for the state
or possibly states surrounding the SMSA's. Each student can build a life
table for his or her SMSA. The study of population dynamics must be
paralleled by analysis of actual data.

The mere phrasing of problems so that they are clear is a distant goal,
hardly to be attained on the first round. Repeated attempts at solution by
colleagues and students, and innumerable justified complaints of obscurity,
are milestones that have to be passed on the way to clarity. Testing has been
possible for many but not all of the problems in this collection. Suggestions
for increased precision in the asking of the questions, as well as for rigor in
the answers, will be received with gratitude.

The great majority of the questions were devised by ourselves, as well as
all of the answers. But our debts are many. Generations of students turned
up ambiguities of wording in their attempts to answer the questions. We
single out for special mention Michael Stoto, Noreen Goldman, Ronald D.
Lee, Juan Carlos Lerda, Robert Lundy, David P. Smith, Edmund M.
Murphy, Peter Ellison, Jay Palmore, and John Hsieh; others at Chicago,
Berkeley, and Harvard participated. The mathematical demography class of
1981–82 at Ball State University worked through many of the chapters. Parts
of the work on the life table in Chapter II were inspired by the examinations
of the Society of Actuaries. Thomas Wonnacott (University of Western
Ontario) used parts of the manuscript in a class, and provided improvements
to several questions. Robert L. Brown (University of Waterloo) and Elias
S.W. Shiu (University of Manitoba) provided corrections for several questions in the Second Printing. Questions that bear on forecasting developed
out of a National Science Foundation grant, whose purpose was to improve
forecasting techniques. Ball State University provided a sabbatical to John
Beekman, in the course of which he made the major part of his contribution
to this joint work. Nathan Keyfitz was greatly helped by the facilities of the
Harvard Center for Population Studies and the Sociology Department of
Ohio State University.

A number of the questions—though none of the answers—were first
published in the revised version of the *Introduction to the Mathematics of
Population*, and we are grateful to Addison-Wesley Publishing Company,
Reading, Massachusetts, for permission to use these. That book is a source
of theory for the reader who wants to go further into the mathematics.
A source of a different kind, that emphasizes applications, is *Applied Mathematical Demography* (John Wiley & Sons, 1977), in which many of the
ideas that are here expressed as problems will be found.

Finally, the typing of successive drafts of the manuscript was the work
of Beatrice Keyfitz and Diane Sorlie.

Contents

CHAPTER I
Populations That Are Not Age-Dependent

Demographers, actuaries, and vital statisticians, like others who use empirical materials, require models for the interpretation of their data. Their work begins with simple assumptions, necessarily highly abstract, and then proceeds to more realistic models that cannot avoid being more complex. This book starts with problems in which age, as well as most other features of real populations, is disregarded, and in which outcomes are taken as deterministic rather than stochastic. A brief introduction presents some of the mathematical concepts of this first preliminary chapter and defines the notation in which the problems are expressed.

1.1. Mathematical Concepts

A population count at time t will be denoted by P_t when discrete time values are being used, or $P(t)$ over a continuum of time. A rate of increase x in population can be expressed as

$$x = \frac{P_{t+1} - P_t}{P_t} \quad \text{or} \quad P_{t+1} = P_t(1 + x) \quad \text{or as} \quad x = \frac{1}{P(t)} \frac{dP(t)}{dt}.$$

Either rate of change can be thought of as a rate of compound interest on a loan.

A population model reflects geometric growth if

$$P_t = P_0(1 + x)^t,$$

where the unit of time may be a month, a year, 5 years, etc., provided x is a fraction of increase per unit of time. If one now uses 1 year as unit of time, but assumes that x is compounded j times per year, then at the end of 1 year

$$P_{t+1} = P_t\left(1 + \frac{x}{j}\right)^j.$$

When the rate x is compounded continuously, it will be called r, and since

$$\lim_{j \to \infty} \left(1 + \frac{r}{j}\right)^j = e^r,$$

we have
$$P_t = P_0 e^{rt}.$$

The rate r is sometimes taken as the crude rate of natural increase:

$$r = \frac{B}{P} - \frac{D}{P},$$

where B and D refer to births and deaths in a year, and P is the population at midyear. The word "crude" reflects the disregard of age.

The projection model $P(t) = Ce^{rt}$ can also be expressed through the differential equation

$$\frac{1}{P(t)} \frac{dP(t)}{dt} = r$$

subject to the initial condition $P(0) = C$. That model implies geometric increase, a growth pattern impossible for more than a limited period. If one allows the derivative to slow down in its growth, one is led to a logistic, i.e.,

$$\frac{dP(t)}{dt} = rP(t)\left[1 - \frac{P(t)}{a}\right].$$

The "slowing down" is effected by assuming an ultimate population, i.e.,

$$\lim_{t \to \infty} P(t) = a.$$

The solution of this differential equation is

$$P(t) = \frac{a}{1 + e^{-r(t - t_0)}},$$

where t_0 is the abscissa of the midpoint of the curve.

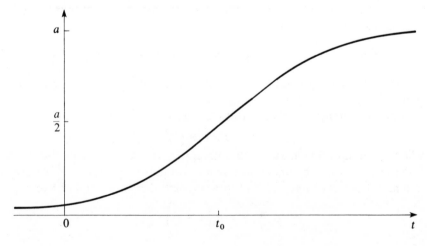

The logistic curve as a population model was first proposed by P. F. Verhulst in 1838 in the paper: "Notice sur la loi que la population suit dans son accroissement," *Correspondence Mathematique et Physique Publiée par A. Quetelet* (Brussels) Vol. 10, pp. 113–121. The logistic curve can be variously described. Thus, several of the problems will use the notation

$$P(t) = [A + Be^{-ut}]^{-1},$$

or

$$P(t) = \frac{1/A}{1 + (B/A)e^{-ut}}.$$

If censuses of p_1, p_2, and p_3 are available at equidistant times t_1, t_2, and $t_3 = (2t_2 - t_1)$, with $p_1 < p_2 < p_3$, then the population ceiling a is given by

$$a = \frac{\dfrac{1}{p_1} + \dfrac{1}{p_3} - \dfrac{2}{p_2}}{\dfrac{1}{p_1 p_3} - \dfrac{1}{p_2^2}},$$

provided the numerator and the denominator in the expression for a are both positive. This is proved in Chapter V, Problem 4.

Populations can be changing with a variable increase $r(t)$. The differential equation

$$r(t) = \frac{1}{P(t)} \frac{dP(t)}{dt},$$

or

$$r(t)\, dt = \frac{dP(t)}{P(t)}$$

describes the growth pattern.

This yields

$$\int_0^T r(t)\, dt = \ln P(t) \Big|_0^T,$$

and by the definition of a logarithm

$$P(T) = P(0) \exp\left[\int_0^T r(t)\, dt\right].$$

A special case is the life-table survivorship in (1) of Chapter II.

Matrices and graphs are useful in demographic problems concerned with transitions among subpopulations. Consider two subpopulations of sizes p_t and q_t at time t.

Let growth and migration be reflected through the equations

$$p_{t+1} = r_{11}p_t + r_{12}q_t,$$

$$q_{t+1} = r_{21}p_t + r_{22}q_t.$$

This can be rewritten with vectors and a matrix as

$$\begin{pmatrix} p_{t+1} \\ q_{t+1} \end{pmatrix} = \begin{pmatrix} r_{11} & r_{12} \\ r_{21} & r_{22} \end{pmatrix} \begin{pmatrix} p_t \\ q_t \end{pmatrix}$$

or more compactly as $P_{t+1} = RP_t$.

If the rates are fixed over time, the population at time $t + n$ equals that at time t successively operated on by R for n times:

$$P_{t+n} = R(\cdots R(RP_t)\cdots) = R^n P_t.$$

Let $\lambda = \lim_{t\to\infty} \|P_{t+1}\|/\|P_t\|$, where $\|P_t\| = p_t + q_t$. The quantity λ is the ultimate ratio of population increase. It can be obtained by finding the dominant root of the equation

$$\begin{vmatrix} r_{11} - \lambda & r_{12} \\ r_{21} & r_{22} - \lambda \end{vmatrix} = 0.$$

Not only does the ratio of total populations tend to stabilize, but the ultimate ratio of subpopulations stabilizes. Thus, there is a constant λ such that population q comes to be λ times as large as population p. A third aspect of stability is that the ultimate ratio of total increase, and ultimate ratio of subpopulations are not affected by the starting values for the subpopulations.

When the number of subgroups of a population is large, the matrices and conditions for stability are very difficult to handle. Luckily, there are ideas in graph theory which can be utilized to provide rules for stability. A graph consists of vertices, corresponding in our problems to subpopulations and edges representing transitions among the subgroups. Our graphs are really digraphs as the edges will always be directed, the directions being indicated by arrows. Any digraph has a corresponding matrix consisting of zeros and ones. Consider the 3×3 matrix M and an initial population vector P:

$$M = \begin{pmatrix} 0 & 1 & 0 \\ 1 & 0 & 1 \\ 1 & 0 & 1 \end{pmatrix} \quad \text{and} \quad P = \begin{pmatrix} p_1 \\ p_2 \\ p_3 \end{pmatrix}.$$

A positive entry m_{ij} (row i, column j) represents a transition from state j to state i; if $i = j$, it represents a transition from state i to itself. A zero entry for m_{ij} reflects no transition from state j to state i. The matrix M is equivalent to the following digraph in which p_1, p_2, and p_3 are the nodes:

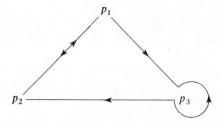

A graph or its corresponding matrix is called *irreducible* if every point or state of the graph can be reached from every other state. An irreducible graph is *primitive* if there exists a number R so that any state can be reached from any other state in exactly R moves; this implies that the corresponding matrix raised to the Rth power contains only positive elements. Primitivity is a condition for stability to occur at high powers of the matrix. A graph is primitive if and only if it contains at least two circuits whose lengths have no common divisor greater than unity. Suggestions for obtaining powers of matrices are given in Solution 17, Chapter V.

1.2. Problems

1. (a) Suppose a population numbering 100,000 growing *geometrically* at annual rate $r = 0.02$. If r is compounded annually—i.e., the population P_1 at the end of 1 year equals $P_0(1 + r)$—determine the size of the population at the end of 10 years.

(b) If another population also of 100,000 is growing *arithmetically* with an increment of 0.02 of its initial number per year, how does its size at the end of 10 years compare with that in (a)?

(c) At the end of 100 years, what is the difference?

2. The official United States population estimate for mid-1965 was 194,303,000; for mid-1970 was 204,879,000. Extrapolate to 1975 assuming (a) fixed absolute increase, i.e., arithmetic progression, and (b) fixed ratio of increase, i.e., geometric progression.

3. (a) A population growing at annual rate r compounded continuously reaches a population at time t equal to $P_t = P_0 e^{rt}$. At what rate r', compounded annually, would it have to grow so as to equal P_t at time t? (Express in terms of r.)

(b) At what rate r_j, where r_j is compounded j times annually, would it have to grow to reach the above population size—i.e., what rate r_j is equivalent to rate r compounded continuously?

4. Obtain recent estimates of the U.S. population. Assume that an exponential growth model can be used for the U.S. population. In how many years will the population grow by 100 million people?

5. With the data and model from Problem 4, in how many years will the population double?

6. A population of 1,000,000 increases at 2 percent per year for 100 years. What difference does it make if the 2 percent is compounded annually or continuously?

7. Prove that compounding j times per year with rate of interest $r + r^2/(2j) = r_j$ is nearly equivalent to a rate of interest r compounded continuously.

8. Doubling time is given by $\ln 2/\ln(1 + r) \doteq 0.693/\ln(1 + r)$, if r is the rate of increase compounded annually. Show

(a) that $0.70/r$ is on the whole a better approximation than $0.693/r$ for values of r of 0.01, 0.02, 0.03, 0.04, and

(b) that for those r values the error in $0.70/r$ is never in excess of one percent.

9. A population of size 5,000,000 closed to migration has fewer births than deaths. If it decreases at an annual rate of 0.0075, how many years will it take to reach half its present size?

10. Suppose that world population numbered 2 persons 1,000,000 years ago and 4,000,000,000 now.

(a) How many doublings does this represent?

(b) What is the average time per doubling?

(c) What is the average rate of exponential increase?

11. Given that 6^{12} is about 2 billion and that $\ln 6$ is 1.79176, work out the average rate of increase in Problem 10 with logarithms to base 6 rather than base 2.

12. A population growing exponentially stood at 1,000,000 in 1930 and at 3,000,000 in 1970.

(a) What is its rate of increase?

(b) What is its doubling time?

(c) How long would it take to multiply by 9?

13. Consider a logistic population model: $P(t) = [A + Be^{-ut}]^{-1}$, $t \geq 0$. Verify that $P(t)$ satisfies the differential equation

$$\frac{dP(t)}{dt} = uP(t) - uA[P(t)]^2.$$

14. Find the coordinates of the point of inflection for $P(t)$.

15. Fit a logistic curve to the populations of Michigan from 1810 to 1970 as found in Census Bureau publications. Which volume or volumes did you use?

16. You have decided to project the United States population by a logistic curve

$$P(t) = \frac{1}{A + Be^{-ut}}, \qquad t \geq 0.$$

You know the populations on July 1, 1979, and July 1, 1980, are 221.1 and 223.9 millions. Determine A, B, and u for low, medium, and high growth patterns leading to year 2076 populations of 350, 450, and 550 millions. As an approximation, you may replace $P(\infty)$ by $P(2076)$. How satisfactory are such projections?

17. Use the volume(s) from Problem 15 to do the following project. Set up a differential equation of the form $dy/dt = uy$ for a Standard Metropolitan Statistical Area (SMSA) which interests you. Let k = population in 1950.

Use the data for 1960 and 1970 to determine u. Determine the appropriate units for t. Solve for y.

Note: The following maps could be consulted in choosing a SMSA:

(i) Standard Metropolitan Statistical Areas, United States Maps, GE-50, No. 55;
(ii) Population Distribution, Urban and Rural, in the United States: 1970, United States Maps, GE-50, No. 45;
(iii) Population Distribution, Urban and Rural, in the United States: 1970, United States Maps, GE-70, No. 1.

Many university libraries have Map Rooms, and the study of the above maps and others can be both enjoyable and enlightening.

18. This problem is based on the pamphlets "Current Population Reports of the United States Bureau of the Census." Consider a differential equation of the form $dy/dt = uy$ for Hawaii. Use the data for July 1, 1977 and July 1, 1978 to determine the growth rate u. Here $t = 1$ corresponds to 1 year from July 1, 1977. Project the population to 1990.

19. Repeat Problem 18 for a state of your choice.

20. A certain country had a population on July 1, 1980 of 15,000,000. The births and deaths during 1980 were 750,000 and 300,000, respectively.

(a) What is the annual crude rate of growth?

(b) Use an exponential model to project the population to 2030 (July 1).

(c) It is thought that the "ultimate" population will be 145,000,000. Set up a logistic curve for the projections.

(d) What is the first year in which a population of 144,500,000 occurs?

21. Consult pages 254–256 of *World Population 1979*, U.S. Dept. of Commerce, Bureau of the Census, 1980. These pages provide population data for Thailand. Verify the annual rates of growth for 1975 and 1978.

With the 1978 rate of growth, and an exponential model, obtain low and high population projections to July 1, 2029.

22. As an alternate model for Problem 21, use a logistic

$$P(t) = \frac{1/A}{1 + (B/A)e^{-ut}}, \qquad t \geq 0,$$

where $P(0) = 46{,}687{,}000$, $P(\infty) = 1/A = 245{,}000{,}000$, and an assumption that $u = 0.023$. What value of B/A results?

Determine $P(10)$ (population on July 1, 1989), $P(15)$, $P(20)$, $P(25)$, $P(30)$, $P(35)$, $P(40)$, and $P(50)$ (population on July 1, 2029).

23. Recall that for the logistic of Problem 22,

$$\frac{1}{P(t)} \frac{dP(t)}{dt} = u - uAP(t).$$

Use the right-hand side of this equation to obtain the relative rates of change for $t = 15, 25, 35$ in Problem 22.

24. Use the model in Problem 22 to determine the first year in which

$$P(t) = \left(\frac{1}{A} - 500{,}000\right) \text{ people.}$$

25. Consult pages 164 and 165 of the reference in Problem 21. Study notes 5 and 11, and obtain a population projection for Zimbabwe to July 1, 2000.

26. Choose a different country from the same reference. Repeat Problems 21 through 24 for it.

27. Write down the simple formula for tripling time at rate of increase r percent per year compounded continuously.

28. If a population is increasing at 2 percent per year, what is the ratio of growth during the lifetime of a person who lives 70 years? At rate r, prove that the ratio of growth averaged over the lifetimes of several members of a population who live an expected $\mathring{e}_0$ years, with standard deviation of length of life equal to σ, is

$$[\exp(r\mathring{e}_0)]\left(1 + \frac{r^2\sigma^2}{2}\right)$$

up to the term in r^2. If the third moment about the mean is 3000, and r is 0.02, what is the error of this expression? (Statistical moments are defined in Chapter VI.)

29. There were 1000 births in a population in the year 1600 and 50,000 in the year 1900. If births were growing exponentially during the period with a constant rate of increase, how many people lived over the period?

30. If the annual births at time t_1 were n_1, and at time t_2 were n_2, what total number of births would occur between t_1 and t_2
  (a) on the assumption of straight line or arithmetic increase, and
  (b) on the assumption of exponential or geometric increase?
  (c) Express the difference between the arithmetic and geometric results as a function of $t_2 - t_1$.

31. (a) If a population starts at one birth per year and grows steadily at rate r, show that by time t, if rt is large, the number of births is e^{rt} and the number that ever lived is nearly e^{rt}/r. Show also that the fraction still alive is close to r/b, where b is the annual birth rate.
  (b) With an annual birth rate of 35 per thousand and a doubling in 10,000 years, what fraction is alive at any given time? With the same birth rate, if the population doubles in 100 years, what fraction is alive at any given time?

32. One thousand immigrants enter a territory at the end of each year, so that $1000t$ enter in t years. If the country was initially uninhabited, and the immigrants increase at rate r, what is the total living population, i.e., surviving immigrants plus native-born descendants, after t years? For what value of t is the total living population double the number of immigrants to date? (Leave your answer in the form of an equation.)

33. Suppose also that the death rate in the previous problem was d, similarly compounded momently. When would the living native-born exceed the surviving immigrants?

34. If a population is growing in such a way that the ratio of current births to persons who ever lived is always the same positive constant f, show that it must be growing geometrically.

35. (a) Consider a population containing two subgroups each initially of size 100,000. If the two subgroups grow exponentially at annual rates of 0.02 and 0.04, respectively, determine the total population after 5 years. What is the average rate of increase $r(t)$ at this time?

(b) Suppose now that the two subgroups are not distinguished—i.e., we have a population initially of size 200,000 growing at an annual rate of 0.03. How does the size of this population at time t compare with that in (a)? Calculate for $t = 5$ years.

36. A population starting with 1,000,000 people was growing exponentially at 3.5 percent per year when it first came under observation; its rate of growth immediately started to fall by the amount of 0.1 percent per year. What was the population at the end of 15 years?

37. The rate of increase of a population at time t is $r(t) = 0.01 + 0.0001t^2$. If the population totals 1,000,000 at time zero, what is it at time 30?

38. A population grows as $P(t) = (\alpha/(t_0 - t))^2$. What is its rate of increase at time t? What does this say about the birthrate as $t \to t_0$?

39. A population increases at an increasing rate; in year t its rate is $r(t) = 0.02 + 10^{-6}t^3$. It starts at $P_0 = 1,000,000$. What is it at time 30? At time t?

40. Fit the modified hyperbola $P(t) = (\alpha/(t_0 - t))^2$ to the population of a hypothetical country:

$$P_{1950} = 16,000,000,$$
$$P_{1970} = 25,000,000.$$

What ominous conclusion for the future of that country do you draw?

41. Given in addition that $P_{1930} = 6,000,000$, fit the logistic

$$P(t) = \frac{a}{1 + \exp(-c(t - t_0))}$$

to the data in Problem 40. Express (a) the initial rate of increase, (b) the ultimate population, and (c) the time at which half the ultimate population is reached, both in terms of the parameters of $P(t)$ and numerically.

42. Consider a population $P(t)$ where $dP(t)/dt$ equals 0.01 times the size the population has already attained less 5×10^{-9} times the square of the size it has attained. Given that the population starts at 1,000,000, set up and solve the equation for the population at time t. What is the ultimate upper limit on its size? Does it depend on the starting population?

43. Use graphs and matrix algebra to determine whether the following matrices are: (a) irreducible; (b) primitive:

$$
\begin{bmatrix} 1 & 0 & 0 & 1 \\ 0 & 0 & 1 & 0 \\ 0 & 1 & 0 & 0 \\ 0 & 0 & 1 & 1 \end{bmatrix}
\quad
\begin{bmatrix} 1 & 1 & 0 & 1 \\ 0 & 0 & 1 & 1 \\ 1 & 0 & 0 & 0 \\ 0 & 1 & 0 & 0 \end{bmatrix}
\quad
\begin{bmatrix} 0 & 1 & 1 & 0 \\ 0 & 0 & 0 & 1 \\ 1 & 0 & 1 & 0 \\ 0 & 1 & 0 & 1 \end{bmatrix}.
$$

44. Consider the aging process below: the population is divided into six age groups, with fertility occurring in only the fourth age group. Is the graph irreducible? What modifications would make the graph primitive?

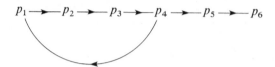

45. Draw a graph that is irreducible but not primitive. Write out the corresponding matrix. How can your graph and matrix be modified so that they will be primitive?

46. Consider a population composed of two subgroups acted on by the 2×2 projection matrix:

$$
\mathbf{R} = \begin{bmatrix} r_{11} & r_{12} \\ r_{21} & r_{22} \end{bmatrix}.
$$

Show that if all the entries in $\mathbf{R}$ are positive, then the ultimate ratio of increase λ is larger than either r_{11} or r_{22}.

47. Show with a graph that if either r_{12} or r_{21} in matrix $\mathbf{R}$ of Problem 46 equals zero, then $\mathbf{R}$ is not irreducible.

48. Each year the net rate of increase of the rural population is 2 percent, of the urban population 1 percent; at the end of each year 0.1 percent of the urban population moves to rural parts, 0.9 percent of the rural to urban.

(a) Set up the transition matrix, and show that it is both irreducible and primitive.

(b) Project through one year, starting with 1,000,000 persons in each population.

(c) What is the ultimate ratio of increase?

(d) What is the ultimate ratio of rural to urban population?

49. In a certain population people are classified as well or (seriously) sick. Each month new entrants (all of them well) are 0.007 of the initially well people; 0.0002 of the people initially classified as well die; 0.01 of those initially well become sick; 0.005 of those initially sick recover; 0.001 of those initially sick die. Set down the equations of the process. What is the stability property of the process?

50. (a) Draw the graphs corresponding to

$$\begin{bmatrix} 1 & 0 & 0 & 1 \\ 1 & 0 & 0 & 1 \\ 1 & 1 & 0 & 0 \\ 0 & 0 & 0 & 1 \end{bmatrix} \quad \begin{bmatrix} 1 & 1 & 0 & 0 \\ 1 & 1 & 0 & 0 \\ 0 & 0 & 1 & 1 \\ 0 & 0 & 1 & 1 \end{bmatrix} \quad \begin{bmatrix} 0 & 0 & 0 & 1 \\ 0 & 0 & 1 & 0 \\ 0 & 1 & 0 & 0 \\ 1 & 0 & 0 & 0 \end{bmatrix}$$

(b) Write the matrix corresponding to

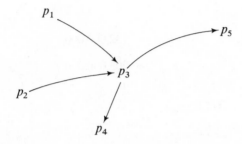

51. Migration at fixed rates takes place from state A to each of B, C, D and E, from B to C, D, and E, and from C to D and E. Show that stability will never be reached.

52. Add the smallest number of elements that make

$$\begin{bmatrix} 0 & 0 & 0 & 0 \\ 1 & 0 & 0 & 0 \\ 1 & 1 & 0 & 1 \\ 0 & 0 & 0 & 0 \end{bmatrix}$$

irreducible.

53. Square the modified matrix of Problem 52 and portray your answer as a graph.

54. What is the smallest number of additional transitions that would make

$$\begin{bmatrix} 0 & 0 & 0 & 0 & 0 & 0 \\ 0 & 0 & 0 & 0 & 1 & 0 \\ 0 & 1 & 0 & 0 & 0 & 0 \\ 1 & 0 & 0 & 0 & 0 & 0 \\ 0 & 0 & 1 & 0 & 0 & 0 \\ 0 & 0 & 0 & 0 & 0 & 0 \end{bmatrix}$$

both irreducible and primitive?

55. In a large country the number of engineers has been increasing for a long time at 4 percent per year, this being the net of entrants on the one hand and retirements and deaths on the other. Each year the new entrants are

10 percent of the number present. What fraction of the engineers who have ever existed are alive and practicing at any given moment?

56. Supposing that the United States population will ultimately be asymptotic to a ceiling of 300,000,000, find the population P_{2000} by using the logistic

$$P_t = \frac{300,000,000}{1 + e^{c(t_0 - t)}}$$

given

$$P_{1940} = 132,000,000,$$

$$P_{1970} = 203,000,000.$$

57. Fit

$$P_t = \frac{\alpha}{t_e - t}$$

to the values of P_{1940} and P_{1970} in the preceding problem. Compare with the logistic extrapolation. The symbol t_e is a time such that $P(t_e) = +\infty$.

58. Prove that projection of P_1 at exponential rate r_1 added to projection of P_2 at exponential rate r_2 gives a higher result than projection of $P_1 + P_2$ at exponential rate

$$r = \frac{P_1 r_1 + P_2 r_2}{P_1 + P_2}.$$

1.3. Solutions

1. (a) $P_{10} = P_0(1 + r)^{10}$
   $= 100,000(1.02)^{10} = 121,899.$
(b) $P_{10} = P_0(1 + 10r)$
   $= 100,000(1.20) = 120,000.$
(c) At the end of 100 years, the difference has grown to 424,465.

2. (a) $P_n = P_0(1 + nr)$ where P_0 is the mid-1965 population.
   $P_5 = P_0(1 + 5r),$
   $1 + 5r = P_5/P_0 \Rightarrow r = 0.010886,$
   $P_{10} = P_0(1 + 10r) = 215,455,000.$
(b) $P_5 = P_0(1 + r)^5,$
   $P_{10} = P_5(1 + r)^5 = 216,031,000.$

3. (a) $P_0 e^{rt} = P_0(1 + r')^t,$
   $r' = e^r - 1.$

(b) $P_0 e^{rt} = P_0 (1 + r')^t = P_0 \left(1 + \dfrac{r_j}{j}\right)^{jt},$

$e^{r/j} = 1 + \dfrac{r_j}{j},$

$r_j = j(e^{r/j} - 1).$

6. $1{,}000{,}000(1.02)^{100} = 7{,}245{,}000,$
   $1{,}000{,}000 e^2 = 7{,}389{,}000.$

7. $e^r = \left(1 + \dfrac{r_j}{j}\right)^j,$

$r_j = j(e^{r/j} - 1)$

$= j\left\{1 + \dfrac{r}{j} + \dfrac{r^2}{2j^2} + \dfrac{r^3}{6j^3} + \cdots - 1\right\}$

$\doteq r + \dfrac{r^2}{2j} + \dfrac{r^3}{6j^2} + \cdots.$

8. $P_t = P_0 (1 + r)^t = 2P_0,$
   $\qquad (1 + r)^t = 2,$
   $\qquad t \ln_e(1 + r) = \ln_e 2.$

The true doubling time for r compounded annually is

$$\frac{\ln 2}{\ln(1 + r)} = 0.693/(r - \frac{r^2}{2} + \frac{r^3}{3} - \cdots)$$

$$\doteq \frac{0.693}{r}\left(1 + \frac{r}{2} - \frac{r^2}{12}\right),$$

of which the term in r^2 is negligible for the rates of increase possible in human populations. The true doubling time being about $(0.693/r)(1 + (r/2)$, we come closer with $0.70/r$ than with $0.693/r$ for $r \geq 0.01$, and the error of $0.70/r$ is

$$\frac{0.70}{r} - \frac{0.693}{r}\left(1 + \frac{r}{2}\right) = \frac{0.007}{r} - \frac{0.693}{2}.$$

Its relative error is

$$\frac{\dfrac{0.007}{r} - \dfrac{0.693}{2}}{\dfrac{0.693}{r}\left(1 + \dfrac{r}{2}\right)} \doteq 0.01 - \frac{r}{2},$$

or $(1 - 50r)$ percent. Thus $0.70/r$ has a relative error of 1 percent near $r = 0$, declining to 0 at $r = 0.02$, rising to 1 percent at $r = 0.04$, and surpassing 1 percent above $r = 0.04$.

9. $P_t = P_0 (1 - 0.0075)^t,$

$$0.5 = 0.9925^t \Rightarrow t = \frac{\ln 0.5}{\ln 0.9925} = 92.08.$$

10. (a) $4,000,000,000 = 2 \cdot 2^n \Rightarrow n = 30.897$ doublings.

(b) $1,000,000/n = 32,365$ years.

(c) $4,000,000,000 = 2e^{r1,000,000} \Rightarrow r = 0.0000215$.

11. $4,000,000,000 = 2e^{r1,000,000}$,

$$6^{12} \doteq e^{r1,000,000},$$

$$12 = 1,000,000r \log_6 e$$

$$= 1,000,000r \frac{\ln e}{\ln 6},$$

$$r = \frac{12(\ln 6)}{1,000,000} = 0.0000215.$$

12. (a) $3,000,000 = 1,000,000e^{40r} \Rightarrow r = 0.027465$.

(b) $2 = e^{0.027465t} \Rightarrow t = 25.24$ years.

(c) 80 years.

14. Using the definition or Problem 13,

$$P'(t) = uP(t) - uAP^2(t)$$

$$\frac{d^2 P(t)}{dt^2} = uP'(t) - 2uAP(t)P'(t) = uP'(t)(1 - 2AP(t)).$$

At $d^2 P(t)/dt^2 = 0$ we have $P(t) = 1/2A$, and this is the ordinate at the point of inflection. The abscissa is the value of t found from

$$\frac{1}{2A} = \frac{1}{A + Be^{-ut}},$$

so

$$t = \frac{-\ln(A/B)}{u}.$$

20. (a) $u = 0.03$.

(b) $P(50) = 67,225,000$.

(c) $P(t) = \{0.0068966 + 0.0597701e^{-0.03t}\}^{-1}$.

(d) $144.5 = P(t) \Rightarrow t = 261$. Thus the population of 144,500,000 people would be achieved in the year $1980 + 261 = 2241$.

21. Low: $46,687,000e^{0.021(50)} = 133,415,000$.

High: $46,687,000e^{0.025(50)} = 162,954,000$.

22.

$$\frac{1}{A + B} = 46,687,000 \Rightarrow \frac{B}{A} = 4.248.$$

$$P(t) = \frac{245,000,000}{1 + 4.248e^{-0.023t}}, \qquad t \geq 0.$$

$P(10) = 55{,}998{,}000;$ $P(15) = 61{,}120{,}000;$ $P(20) = 66{,}545{,}000;$

$P(25) = 72{,}263{,}000;$ $P(30) = 78{,}257{,}000;$ $P(35) = 84{,}505{,}000;$

$P(40) = 90{,}980{,}000;$ $P(50) = 104{,}474{,}000.$

23. $t = 15 \rightarrow 0.023(1 - 0.249469) = 0.017,$
$t = 25 \rightarrow 0.023(1 - 0.294953) = 0.016,$
$t = 35 \rightarrow 0.023(1 - 0.344919) = 0.015.$

24.

$$244{,}500{,}000 = \frac{245{,}000{,}000}{1 + 4.248e^{-0.023t}} \quad \text{yields } t = 332.$$

Hence $1979 + 332 = 2311$ would be the approximate year.

25. Such a projection would begin with the July 1, 1979 population, and multiply by appropriate growth factors $1 + r_i$, $i = 1, 2, \ldots, 21$. The factors would begin with low numbers because of war losses.

27. $3 = e^{rt} \Rightarrow t = \ln 3/r \doteq 1.1/r.$

28. $\dfrac{P_{t+1}}{P_t} = e^{0.02} =$ ratio of growth of population during tth year.

$\dfrac{P_{t+70}}{P_t} = e^{1.40} = 4.055 =$ ratio of growth of population during those 70 years.

Call the length of life of the ith individual x_i, $i = 1, 2, \ldots, N$; then the ratio of increase during his life-time is e^{rx_i}, and one needs the average $\sum_i e^{rx_i}/N$ or

$$\frac{1}{N} \sum_i e^{rx_i} = \frac{1}{N} \exp(r\mathring{e}_0) \sum_i \exp(r(x_i - \mathring{e}_0))$$

$$= \frac{1}{N} \exp(r\mathring{e}_0) \sum_i \left[(1 + r(x_i - \mathring{e}_0) + \frac{r^2}{2}(x_i - \mathring{e}_0)^2 + \cdots \right]$$

$$= \exp(r\mathring{e}_0)\left[1 + \frac{r^2\sigma^2}{2} + \frac{r^3\mu_3}{6} + \cdots \right],$$

where μ_3 is the third moment about the mean of length of life.

Error in the expression $= e^{0.02\mathring{e}_0} \dfrac{(0.02)^3(3000)}{6} = 0.004e^{0.02\mathring{e}_0}.$

29. $50{,}000 = 1000e^{300r} \Rightarrow r = 0.01304$

$$\text{Total births} = \int_0^{300} B(t)dt$$

$$= \int_0^{300} 1000e^{\prime 0.1304t}\, dt.$$

Total people who lived $= P_0 + 3{,}758{,}000.$

30. (a)

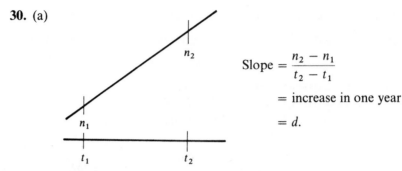

$$\text{Slope} = \frac{n_2 - n_1}{t_2 - t_1}$$

$$= \text{increase in one year}$$

$$= d.$$

Total number of births

$$= \int_{t_1}^{t_2} \{n_1 + \frac{n_2 - n_1}{t_2 - t_1} (t - t_1)\} dt$$

$$= \left(\frac{n_1 + n_2}{2}\right)(t_2 - t_1)$$

31. (a) $B_t = B_0 e^{rt} = e^{rt}$.
The total number who ever lived is

$$\int_0^t e^{rs} \, ds = (e^{rt} - 1)/r$$

$$\doteq e^{rt}/r.$$

If b is the annual birth rate,

$$b = \frac{\text{number births at time } t}{\text{number alive at time } t} = \frac{e^{rt}}{\text{number alive at time } t}.$$

Therefore,

$$\text{"Fraction still alive"} = \frac{e^{rt}/b}{e^{rt}/r} = \frac{r}{b}.$$

(b) $b = 0.035$,
$\quad 2 = e^{r10,000} \Rightarrow r = 0.0000693147$, and $r/b = 0.00198$.
If $\quad 2 = e^{r100}$, then $r = 0.00693147$, and $r/b = 0.198$.

33. At the end of t years surviving immigrants would be

$$I = 1000 \frac{1 - e^{-dt}}{1 - e^{-d}}.$$

Native-born would be the total L of the preceding problem less this: $L - I$, and the native-born would exceed the immigrants when

$$L - I > I \quad \text{or} \quad L > 2I \quad \text{or} \quad \frac{e^{rt} - 1}{e^r - 1} > 2\frac{1 - e^{-dt}}{1 - e^{-d}}.$$

Numerical work shows for a country peopled by immigration that the native-born quickly come to outnumber the newcomers when r is large.

34. $B(t) = f \int_0^t B(u) \, du,$

$\dfrac{dB(t)}{dt} = f B(t).$ Hence, $B(t)$ is an exponential function.

35. (a) $100{,}000 e^{0.1} = 110{,}517,$

$$100{,}000 e^{0.2} = \frac{122{,}140}{232{,}657},$$

$$232{,}657 = 200{,}000 e^{5r} \Rightarrow r = 0.03025.$$

(b) $200{,}000 e^{5(0.03)} = 232{,}367.$

36.

$$\frac{1}{P(t)} \frac{dP(t)}{dt} = 0.035 - 0.001t = r(t),$$

$$P(t) = P_0 e^{\int_0^t r(w) \, dw},$$

$$P(15) = 1{,}000{,}000 e^{\int_0^{15} [0.035 - 0.001w] \, dw}$$

$$= 1{,}510{,}590.$$

37.

$$P(30) = 1{,}000{,}000 e^{\int_0^{30} [0.01 + 0.0001w^2] \, dw}$$

$$= 3{,}320{,}117.$$

38.

$$\frac{dP(t)}{dt} = 2\alpha^2 (t_0 - t)^{-3}.$$

The birthrate approaches $+\infty$ as $t \to t_0$.

39.

$$P(30) = 1{,}000{,}000 \exp\left\{ \int_0^{30} [0.02 + 10^{-6} w^3] \, dw \right\}$$

$$= 2{,}231{,}112.$$

$$P(t) = 1{,}000{,}000 e^{0.02t + 10^{-6} t^4 / 4}.$$

40. Let $t = 0$ represent 1950, and $t = 20$ represent 1970. The given data imply

$$P(0) = 16 = \left(\frac{\alpha}{t_0} \right)^2, \quad \text{and}$$

$$P(20) = 25 = \left(\frac{\alpha}{t_0 - 20} \right)^2.$$

Taking square roots and solving, one obtains $t_0 = 100$, and $\alpha = 400$. Thus,

$$P(t) = \left(\frac{400}{100 - t} \right)^2.$$

As $t \to 100$ (the year 2050), the population explodes to $+\infty$.

42.

$$\frac{dP(t)}{dt} = 0.01P(t) - 5 \times 10^{-9}[P(t)]^2$$

$$= 0.01P(t)\left[1 - \frac{5}{10^7}P(t)\right],$$

with $P(0) = 1{,}000{,}000$.

Comparing this with the differential equation

$$\frac{dP(t)}{dt} = rP(t)\left[1 - \frac{P(t)}{a}\right]$$

we have $r = 0.01$ and $a = 2{,}000{,}000$, and at $P(t) = 1{,}000{,}000$, or $a/2$ we have $t_0 = 0$. Thus the required logistic is

$$P(t) = \frac{2{,}000{,}000}{1 + e^{-0.01t}}.$$

The ultimate population is 2,000,000 regardless of where on the curve P_0 is. The asymptote is determined by the differential equation and does not depend on the initial condition that locates the origin on the chart.

43. The first matrix translates to the graph

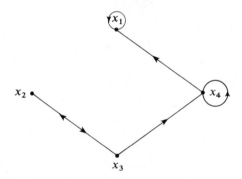

This is not irreducible (cannot leave state 1).

The second matrix translates to the graph

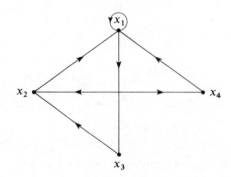

This is irreducible; primitive with $R = 4$. We have $M^4 = MM^3$, or

$$\begin{pmatrix} 1 & 1 & 0 & 1 \\ 0 & 0 & 1 & 1 \\ 1 & 0 & 0 & 0 \\ 0 & 1 & 0 & 0 \end{pmatrix}^4 = \begin{pmatrix} 1 & 1 & 0 & 1 \\ 0 & 0 & 1 & 1 \\ 1 & 0 & 0 & 0 \\ 0 & 1 & 0 & 0 \end{pmatrix} \begin{pmatrix} 2 & 3 & 2 & 3 \\ 1 & 1 & 1 & 2 \\ 1 & 2 & 1 & 2 \\ 1 & 1 & 0 & 0 \end{pmatrix}$$

$$= \begin{pmatrix} 4 & 5 & 3 & 5 \\ 2 & 3 & 1 & 2 \\ 2 & 3 & 2 & 3 \\ 1 & 1 & 1 & 2 \end{pmatrix}.$$

The third matrix converts to the graph

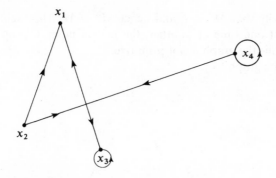

The graph is not irreducible.

44. No; from state 6, the process goes nowhere. If one connects age groups 5 and 6 with age group 1, and drops the connection between age groups 4 and 1, the graph is primitive with $R = 26$. The matrix M, and its eighth and twenty-sixth powers are

$$M = \begin{pmatrix} 0 & 0 & 0 & 0 & 1 & 1 \\ 1 & 0 & 0 & 0 & 0 & 0 \\ 0 & 1 & 0 & 0 & 0 & 0 \\ 0 & 0 & 1 & 0 & 0 & 0 \\ 0 & 0 & 0 & 1 & 0 & 0 \\ 0 & 0 & 0 & 0 & 1 & 0 \end{pmatrix}, \quad M^8 = \begin{pmatrix} 0 & 0 & 1 & 2 & 1 & 0 \\ 0 & 0 & 0 & 1 & 2 & 1 \\ 1 & 0 & 0 & 0 & 1 & 1 \\ 1 & 1 & 0 & 0 & 0 & 0 \\ 0 & 1 & 1 & 0 & 0 & 0 \\ 0 & 0 & 1 & 1 & 0 & 0 \end{pmatrix},$$

$$M^{26} = \begin{pmatrix} 5 & 10 & 10 & 5 & 2 & 1 \\ 1 & 5 & 10 & 10 & 5 & 1 \\ 1 & 1 & 5 & 10 & 10 & 4 \\ 4 & 1 & 1 & 5 & 10 & 6 \\ 6 & 4 & 1 & 1 & 5 & 4 \\ 4 & 6 & 4 & 1 & 1 & 1 \end{pmatrix}.$$

45.

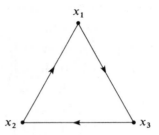

The matrix

$$M = \begin{pmatrix} 0 & 1 & 0 \\ 0 & 0 & 1 \\ 1 & 0 & 0 \end{pmatrix}.$$

One can verify that $M^3 = I$, and hence $M^4 = M$. Although each position is positive at one time or another, the matrix never fills up with positive numbers. Thus the graph is not primitive.

Change the graph to

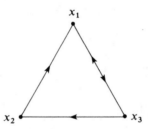

Then

$$M^* = \begin{pmatrix} 0 & 1 & 1 \\ 0 & 0 & 1 \\ 1 & 0 & 0 \end{pmatrix} \quad \text{and} \quad (M^*)^5 = \begin{pmatrix} 2 & 1 & 2 \\ 1 & 1 & 1 \\ 1 & 1 & 2 \end{pmatrix}.$$

46. At stability let the stable vector be (p, q), so that $r_{11}p + r_{12}q = \lambda p$. If r_{12} and q are both positive then $r_{11}p < \lambda p$, therefore, $r_{11} < \lambda$. Similarly, from $r_{21}p + r_{22}q = \lambda q$ we have $r_{22} < \lambda$. (We exclude the trivial case in which either $p = 0$ or $q = 0$ and hence the other equals 0.) The result is easy to see graphically. Since λ is the dominant root of the equation

$$\begin{vmatrix} r_{11} - \lambda & r_{12} \\ r_{21} & r_{22} - \lambda \end{vmatrix} = 0,$$

or $(r_{11} - \lambda)(r_{22} - \lambda) - r_{21}r_{12} = 0$, then $(r_{11} - \lambda)(r_{22} - \lambda)$ being a parabola concave upward, intersecting the abscissa at r_{11} and r_{22}, its intersections with the horizontal line at $r_{21}r_{12}$ must be larger than r_{22} and smaller than $r_{11} < r_{22}$. By considering derivatives we see that the dominant value λ_1 must be at the point marked, and that it must be to the right of both r_{11} and r_{22}.

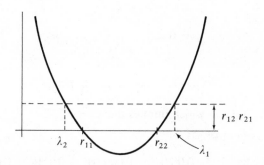

47. If $r_{21} = 0$, the graph is

and passage from x_1 is impossible. Comparable results hold when $r_{12} = 0$.

48. (a) Let $P^{(t+1)} = \begin{pmatrix} _uP^{(t+1)} \\ _rP^{(t+1)} \end{pmatrix}$ be the urban and rural populations at time $t + 1$. The growth matrix $G = \begin{pmatrix} 1.01 & 0 \\ 0 & 1.02 \end{pmatrix}$, and the movement matrix $M = \begin{pmatrix} 0.999 & 0.009 \\ 0.001 & 0.991 \end{pmatrix}$. The transition matrix is

$$T = MG = \begin{pmatrix} 1.00899 & 0.00918 \\ 0.00101 & 1.01082 \end{pmatrix} \doteq \begin{pmatrix} 1.009 & 0.009 \\ 0.001 & 1.011 \end{pmatrix}.$$

Since all entries in T are positive, the graph is irreducible, and primitive. $P^{(t+1)}$ is found from the equation $P^{(t+1)} = TP^{(t)}$.

(b) $P^{(1)} = MG\begin{pmatrix} 1{,}000{,}000 \\ 1{,}000{,}000 \end{pmatrix} = M\begin{pmatrix} 1{,}010{,}000 \\ 1{,}020{,}000 \end{pmatrix} = \begin{pmatrix} 1{,}018{,}170 \\ 1{,}011{,}830 \end{pmatrix}.$

Note the reversal of the larger components.

(c) The quantity λ (ultimate ratio of increase) is the dominant root of the equation

$$\begin{vmatrix} 1.009 - \lambda & 0.009 \\ 0.001 & 1.011 - \lambda \end{vmatrix} = 0 \quad \text{or} \quad \lambda^2 - 2.02\lambda + 1.02 = 0.$$

The dominant root is $\lambda = 1.02$.

(d) For large t,

$$_uP^{(t+1)} = \lambda(_uP^{(t)})$$
$$= 1.02(_uP^{(t)}).$$

For large t, $_rP^{(t)} = c(_uP^{(t)})$ for some constant c.

But

$$_uP^{(t+1)} = r_{11}(_uP^{(t)}) + r_{12}(_rP^{(t)}),$$

or

$$_uP^{(t+1)} = (r_{11} + cr_{12})(_uP^{(t)}).$$

Thus, $\lambda = r_{11} + cr_{12}$, which implies that $c = 1.222$.

49. States of nature: W or S.

$$W_{t+1} = W_t + 0.007W_t - 0.0002W_t - 0.01W_t + 0.005S_t,$$

$$S_{t+1} = S_t + 0.01W_t - 0.005S_t - 0.001S_t$$

$$\begin{pmatrix} W_{t+1} \\ S_{t+1} \end{pmatrix} = \begin{pmatrix} 0.9968 & 0.0050 \\ 0.0100 & 0.9940 \end{pmatrix} \begin{pmatrix} W_t \\ S_t \end{pmatrix}.$$

The matrix is primitive with $R = 1$.

50. (a) The first graph is

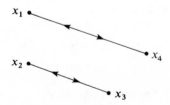

The second graph is

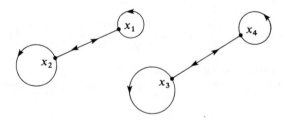

The third graph is

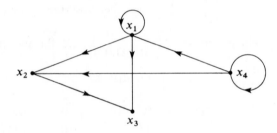

(b) The corresponding matrix is

$$\begin{pmatrix} 0 & 0 & 0 & 0 & 0 \\ 0 & 0 & 0 & 0 & 0 \\ 1 & 1 & 0 & 0 & 0 \\ 0 & 0 & 1 & 0 & 0 \\ 0 & 0 & 1 & 0 & 0 \end{pmatrix}.$$

51. The corresponding matrix is

$$\begin{pmatrix} 0 & 0 & 0 & 0 & 0 \\ 1 & 0 & 0 & 0 & 0 \\ 1 & 1 & 0 & 0 & 0 \\ 1 & 1 & 1 & 0 & 0 \\ 1 & 1 & 1 & 0 & 0 \end{pmatrix} = M.$$

The first row of any power of M will be all zeros, and hence a stable condition will not be reached.

52. The existing matrix is reflected in the graph:

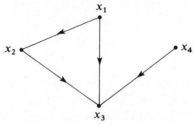

To make this irreducible, one needs to go from x_3 to x_1, and from x_3 to x_4.

Hence, we add the elements $m_{13} = 1$, and $m_{43} = 1$. (There exist other solutions.)

53.

$$\begin{pmatrix} 0 & 0 & 1 & 0 \\ 1 & 0 & 0 & 0 \\ 1 & 1 & 0 & 1 \\ 0 & 0 & 1 & 0 \end{pmatrix}\begin{pmatrix} 0 & 0 & 1 & 0 \\ 1 & 0 & 0 & 0 \\ 1 & 1 & 0 & 1 \\ 0 & 0 & 1 & 0 \end{pmatrix} = \begin{pmatrix} 1 & 1 & 0 & 1 \\ 0 & 0 & 1 & 0 \\ 1 & 0 & 2 & 0 \\ 1 & 1 & 0 & 1 \end{pmatrix}.$$

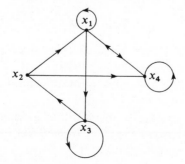

54. The existing matrix is reflected in the graph:

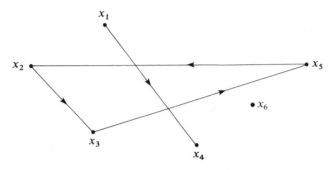

To make this irreducible, we go from x_4 to x_5, from x_5 to x_6, from x_6 to x_1, and from x_1 to x_2.

The following enlarged matrix M is primitive with all of the entries of M^9 being positive.

$$M = \begin{vmatrix} 0 & 0 & 0 & 0 & 0 & 1 \\ 1 & 0 & 0 & 0 & 1 & 0 \\ 0 & 1 & 0 & 0 & 0 & 0 \\ 1 & 0 & 0 & 0 & 0 & 0 \\ 0 & 0 & 1 & 1 & 0 & 0 \\ 0 & 0 & 0 & 0 & 1 & 0 \end{vmatrix}.$$

55. This problem is similar to Problem 31. The growth rate $r = 0.04$, the "birth" rate $b = 0.10$, and hence the fraction of the engineers who are alive and practicing is $0.04/0.10 = 0.4$.

56. If p_1, p_2, and p_3 represent the populations in 1940, 1970, and 2000, we have the formula

$$300,000,000 = \frac{1/p_1 + 1/p_3 - 2/p_2}{1/(p_1 p_3) - 1/p_2^2}.$$

Substituting the given values yields $p_3 = 254,400,000$.

57. The equations

$$132,000,000 = \frac{\alpha}{t_e - 1940} \quad \text{and} \quad 203,000,000 = \frac{\alpha}{t_e - 1970}$$

yield $t_e = 2025.775$. Either equation then gives $\alpha \doteq 11,322,300,000$. This provides the value $P_{2000} = 439,300,000$. This is much higher than the logistic extrapolation, and reflects the population "explosion" built into the hyperbola model in year 2026.

58. Two answers to this problem will be given, based on pages 14–16 of Keyfitz (1977), and pages 12–13 of *Inequalities* by E. F. Beckenbach and R. Bellman, Springer-Verlag, NY, 1965.

At time t, the projections would be $P_1 e^{r_1 t}$, $P_2 e^{r_2 t}$, and $(P_1 + P_2)e^{rt}$. Denote the sum $P_1 e^{r_1 t} + P_2 e^{r_2 t}$ by $P(t)$. Its rate of increase would be

$$\bar{r}(t) = \frac{1}{P(t)} \frac{dP(t)}{dt} = \frac{P_1 r_1 e^{r_1 t} + P_2 r_2 e^{r_2 t}}{P_1 e^{r_1 t} + P_2 e^{r_2 t}}.$$

One can show that

$$\frac{d\bar{r}(t)}{dt} = \frac{P_1 P_2 e^{(r_1 + r_2)t}(r_1 - r_2)^2}{(P_1 e^{r_1 t} + P_2 e^{r_2 t})^2}$$

which is positive, unless $r_1 = r_2$. By contrast,

$$\frac{d}{dt} \left\{ \frac{(P_1 + P_2)re^{rt}}{(P_1 + P_2)e^{rt}} \right\} = 0.$$

Thus the sum of the separate projections is greater.

The second proof relies on the arithmetic-mean–geometric-mean inequality with arbitrary positive weights α, $1 - \alpha$, which is

$$\alpha x_1 + (1 - \alpha)x_2 \geq x_1^{\alpha} x_2^{1-\alpha}, \qquad 0 < \alpha < 1,$$

arbitrary $x_1, x_2 \geq 0$. The equality holds if and only if $x_1 = x_2$. The separate projection of the two populations is $P_1 e^{r_1 t} + P_2 e^{r_2 t}$, whereas the combined projection is $(P_1 + P_2)e^{rt}$. The sum of the separate projections is greater if

$$\frac{P_1}{P_1 + P_2} e^{r_1 t} + \frac{P_2}{P_1 + P_2} e^{r_2 t} > e^{rt}.$$

Letting

$$\alpha = \frac{P_1}{P_1 + P_2}, \qquad x_1 = e^{r_1 t}, \qquad x_2 = e^{r_2 t},$$

in the A.M.–G.M. inequality, we can see that the desired result holds.

CHAPTER II
The Life Table

A life table is a mathematical model which portrays

(a) probabilities of dying at the several ages;
(b) diminution of a cohort as it grows older;
(c) a stationary population.

The construction of a life table begins with the collection and analysis of data. Such population statistics will have been gathered by a national census department, a state vital statistics unit, a life insurance company, a consulting actuarial firm, a medical research team, or possibly a pharmaceutical house. Graduation of data helps remove irregularities produced by errors and randomness. A third step is the conversion of data into appropriate rates, and thence into probabilities. A fourth step is obtaining individual age values for the various functions by interpolation formulas. A fifth step is to use approximate integration techniques to obtain functions of the basic $l(x)$ column.

Life tables are used by actuaries, vital statisticians, and medical researchers to determine life insurance premiums, pension values, gains in life expectancy of a people, and decreased probabilities of death from improved drugs and surgical techniques.

2.1. Mathematical Concepts

The basic life table column is $l(x)$, $0 \leq x \leq \omega$. That function represents the numbers of living at all ages up to the "final" age ω. In some cases, we are only interested in this function at the non-negative integers $x = 0, 1, 2, \ldots$, and the function will be written as l_x. A second column is d_x, $x = 0, 1, 2, \ldots$, the number of deaths between ages x and $x + 1$, i.e., $d_x = l_x - l_{x+1}$. The probability of dying between ages x and $x + 1$ is q_x. Thus $q_x = d_x/l_x$. The function

$$L_x = \int_0^1 l(x + t)\, dt,$$

and its extension is

$$T_x = \int_0^{\omega - x} l(x + t)\, dt.$$

The complete expectation of life at age x is denoted and defined by

$$\mathring{e}_x = \frac{1}{l_x} T_x = \int_0^{\omega - x} \frac{l(x + t)}{l_x}\, dt.$$

The curtate expectation of life

$$e_x = \frac{1}{l_x} \sum_{n=1}^{\infty} l_{x+n}.$$

These two functions are related by the approximate equation $\mathring{e}_x \doteq e_x + 0.5$.

There is another life table function which is most useful but frequently not printed in tables. That is the force of mortality, denoted and defined as follows:

$$\mu(x) = -\frac{1}{l(x)} \frac{dl(x)}{dx}.$$

This is equivalent to

$$\mu(x) = -\frac{d}{dx} \ln_e l(x).$$

After integration and exponentiation, this becomes

$$l(x) = l(0) \exp\left\{ - \int_0^x \mu(y)\, dy \right\}, \qquad (1)$$

a special case of the expression for population increase of Chapter I. The quantity $l(0)$ or l_0 is called the radix of a life table. It is chosen for convenience and may be a number such as 100,000. The probability of living from age x to age $x + n$ is defined by

$$_n p_x = l_{x+n}/l_x.$$

Properties of integrals and exponents allow us to write

$$_n p_x = \exp\left\{ - \int_x^{x+n} \mu(y)\, dy \right\}$$

$$= \exp\left\{ - \int_0^n \mu(x + t)\, dt \right\}.$$

The complementary probability

$$_nq_x = 1 - {}_np_x$$

$$= 1 - \exp\left\{-\int_0^n \mu(x + t)\,dt\right\}.$$

The basic definition of $\mu(x)$ can be rewritten as $l(x)\mu(x) = -(d/dx)l(x)$. If one integrates both sides, uses the fundamental theorem of calculus, and changes variables, one obtains

$$l_x - l_{x+n} = \int_0^n l(x + t)\mu(x + t)\,dt.$$

The probability $_nq_x$ can be obtained by dividing both sides by l_x:

$$_nq_x = \frac{l_x - l_{x+n}}{l_x} = \int_0^n {}_tp_x\mu(x + t)\,dt.$$

A number of people have attempted to describe the $l(x)$ function in an analytic form. One such expression was published in 1860 by W. M. Makeham in the article "On the law of mortality, and the construction of annuity tables," which appeared in Volume 8 of the *Journal of the Institute of Actuaries*. He wrote $\mu_x = A + Bc^x$ to reflect a geometrically growing force coupled with a constant exposure to accidents. Makeham's mortality law built upon the earlier expression $\mu_x = Bc^x$ contained in the 1825 paper by B. Gompertz entitled "On the nature of the function expressive of the law of human mortality," *Philosophical Transactions*, Royal Society of London.

By assuming "the average exhaustion of a man's power to avoid death to be such that at the end of equal infinitely small intervals of time he lost equal portions of his remaining power to oppose destruction which he had at the commencement of these intervals," Gompertz was led to the differential equation

$$\frac{d}{dx}\left(\frac{1}{\mu(x)}\right) = -h\frac{1}{\mu(x)},$$

where h is a constant of proportionality. Upon integration,

$$\ln\left(\frac{1}{\mu(x)}\right) = -hx - \ln B,$$

where $-\ln B$ is a constant of integration. Thus, $\mu_x = e^{hx + \ln B} = Bc^x$, for $c = e^h$.

Makeham's law leads to $l_x = ks^x g^{c^x}$ where $\ln s = -A$, $\ln g = -B/\ln c$, and $k = l_0/g$.

Census records do not produce values of l_x, but rather numbers of people whose age is between x and $x + 1$. Thus, the numbers which are collected relate more to the function

$$L_x = \int_0^1 l(x + t)\, dt$$

for $x = 0, 1, 2, \ldots$. Moreover, the rate d_x/L_x is not q_x. Instead, it is what is called the central death rate at age x and denoted by m_x. Thus, $m_x = d_x/L_x$. If one assumes a uniform distribution of deaths throughout the year of age,

$$l(x + t) = l_x - t\, d_x, \qquad 0 \le t \le 1.$$

Then, $L_x = \int_0^1 l(x + t)\, dt = l_x - \tfrac{1}{2} d_x$,

$$m_x = \frac{q_x}{1 - \tfrac{1}{2} q_x} \quad \text{and} \quad q_x = \frac{m_x}{1 + \tfrac{1}{2} m_x}.$$

This last formula suggests how observed rates can be used to obtain probabilities. The word "suggests" is appropriate because the census people do not observe d_x and L_x, but rather d_x^z and L_x^z where the superscript z denotes a calendar year z. Thus the observed rate d_x^z/L_x^z is denoted by a different symbol, M_x^z. We will drop the z portion of the symbol, but retain the M to distinguish it from the corresponding m rate.

Let the function $p(x)$, $0 \le x \le \omega$, represent the population under study, and the basis for a future life table. The population aged x to $x + n$ is denoted and defined by

$$_nP_x = \int_0^n p(x + t)\, dt.$$

Such a population at time t will be called $_nP_x^{(t)}$.

Most of our previous definitions can be extended to n year intervals. The following definitions will be needed.

$_na_x$ Average number of years lived between ages x and $x + n$ by those dying in that age interval.

$_nd_x$ Number dying in the stationary population between ages x and $x + n$: $_nd_x = l_x - l_{x+n}$.

$_nD_x$ Deaths observed between ages x and $x + n$.

$_nD_x^{(i)}$ Deaths observed between ages x and $x + n$ due to cause i.

$_nL_x$ Number living in life table between ages x and $x + n$: $_nL_x = \int_0^n l(x + t)\, dt$.

$_nM_x$ Observed age-specific death rate for individual between exact ages x and $x + n$.

$_nm_x$ Age-specific death rate in the life table population: $_nm_x = {_nd_x}/{_nL_x}$ $(_1m_x = m_x)$.

$_nq_x$ Probability of dying within n years for individuals now of exact age x: $_nq_x = {_nd_x}/l_x$ $(_1q_x = q_x)$.

The last function should not be confused with $_n|q_x$ which is the probability that a person now aged x will die between ages $x + n$ and $x + n + 1$.

Users of life tables for specific geographic areas obtain valuable information by comparing such rates with those from a standard population, such as a current U.S. Life Table. The symbols

$$_nP^s_x, \; ^m_nP^s_x, \; ^f_nP^s_x, \; _nD^s_x, \; ^m_nD^s_x, \text{ and } ^f_nD^s_x$$

denote total population, male population, female population, total deaths, male deaths, and female deaths from the standard life table for the age range x to $x + n$. To filter out the effects of the population age distribution on the crude death rates, there are adjusted rates. The adjusted death rate by the direct method is

$$\frac{\sum\limits_x {}_nP^s_x \cdot {}_nm_x}{\sum\limits_x {}_nP^s_x}.$$

The adjusted death rate by the indirect method is

$$\frac{D}{\sum\limits_x {}_nm^s_x \cdot {}_nP_x} \cdot \frac{\sum\limits_x {}_nP^s_x \cdot {}_nm^s_x}{\sum\limits_x {}_nP^s_x}.$$

As stated on page 103 of Spiegelman (1968), "the age-adjusted death rate (direct method) is the crude death rate which would be observed in the standard population if it had itself experienced the age-specific death rates of the community or state in question." If the age-specific death rates for the community are not available, the indirect method of adjustment must be followed.

The life table describes a stationary population supported by l_0 annual births distributed uniformly over each calendar year, whose deaths occur in accordance with the given vector $\{q_0, q_1, \ldots, q_{\omega-1}\}$, and for which there is no migration into or out of the population. The word stationary reflects the constant age distribution $\{l_0, l_1, \ldots, l_{\omega-1}\}$ for the population. The assumption of a constant number B of births each year can be generalized by assuming that births follow an exponential function Be^{rt}, $t \geq 0$, where r is the growth rate. This will affect the age distributions in future years. Consider a population closed to migration and subject to a fixed vector $\{q_0, q_1, \ldots, q_{\omega-1}\}$, with births increasing exponentially. These three conditions produce a stable age distribution in which three quantities are increasing exponentially in the same ratio. Those three quantities are the total population, the number of persons living in each age group, and the deaths in each age group. Stable population theory and problems will be treated in Chapter III. One result will be needed now for a few problems. That says that the function $l(x)$ should be replaced by $e^{-rx}l(x)$, $0 \leq x \leq \omega$, to reflect the smaller number of births x years ago. That is, the normal interpretation of l_x as the expected survivors to age x from l_0 births does not apply in a growing population.

The Society of Actuaries' textbook *Actuarial Mathematics* by Bowers, Gerber, Hickman, Jones, and Nesbitt (1983) discusses life tables in Chapter 3. It adopts "the basic viewpoint that a life table is a way of specifying the distribution of the random variable time until death." The random survivorship group is contrasted with the deterministic survivorship group. If one denotes by $L(x)$ the random number of survivors to age x, then $E[L(x)]$ and l_x represent the same quantity. Similarly, if $_nD_x$ denotes the random number of deaths between ages x and $x + n$, then $E[_nD_x]$ and $_nd_x$ have the same value. Readers would profit from a study of life tables in *Actuarial Mathematics*. The problems in this chapter are concerned with a deterministic survivorship group, converting census records to rates to probabilities, adjusting death rates by standard populations, and life table functions.

2.2. Problems

2.2.1. Various Death Rates and Ratios

For Problems 1–5, you have the following hypothetical data set

Indiana

Age group	$_nP_x^s$	$_n^mP_x^s$	$_n^fP_x^s$	$_nD_x^s$	$_n^mD_x^s$	$_n^fD_x^s$
<1	100,000	50,000	50,000	750	400	350
1–4	400,000	200,000	200,000	3,000	1,600	1,400
5–20	1,500,000	750,000	750,000	12,000	6,500	5,500
21–100	4,000,000	1,500,000	2,500,000	31,000	12,000	19,000

Muncie

Age group	$_nP_x$	$_n^mP_x$	$_n^fP_x$	$_nD_x$	$_n^mD_x$	$_n^fD_x$
<1	1,500	750	750	12	6	6
1–4	6,000	3,000	3,000	45	24	21
5–20	20,000	10,000	10,000	160	85	75
21–100	44,000	17,000	27,000	350	200	150

1. Compute the crude death rates for Indiana and Muncie.
2. Compute the 24 $_nm_x$ rates.

3. Regard Indiana as the standard population for Muncie. Compute the adjusted death rate for Muncie by the direct method.

4. Repeat Problem 3 by the indirect method.

5. Compute the age–sex-adjusted rate for Muncie.

6. Derive the expression for Variance(M), where

$$M = \frac{D}{P} = \sum_x \left(\frac{P_x}{P}\right) M_x,$$

if all deaths occur independently.

The next seven problems should use Census Bureau data for your SMSA chosen in Chapter I, Problem 17. Use Table 7-4 of "Vital Statistics of the United States, 1975, Vol. II, Mortality, Part B." Record deaths for the various age groups by sex. Also use "1970 Census of Population, Vol. 1, Characteristics of the Population" for the appropriate state. Chapter B gives numbers alive for various age groups. Use 1960 and 1970 volumes and data to extrapolate to 1975 data.

7. Create central death rates for the various age groups, by sex.

8. Compute crude death rates for 1960, and 1975.

9. Compare the death rates for 1960 and 1975 for your SMSA due to malignant neoplasms versus major cardiovascular diseases.

10. Compute the infant mortality rates for 1960 and 1975.

11. Calculate the adjusted death rate (direct method). The standard population figures are for the state or possibly states surrounding your SMSA.

12. Calculate the age–sex-adjusted death rates.

13. Calculate the adjusted death rate (indirect method).

14. Calculate the variances and standard deviations for the crude death rate, adjusted death rate (direct method), adjusted death rate (indirect method) for your SMSA.

15. Let $X_1, X_2, \ldots, X_n$ be independent random variables, each having distribution

$$X = \begin{cases} 1, & \text{with prob. } q_x, \\ 0, & \text{with prob. } p_x. \end{cases}$$

Define a random variable

$$R = \sum_{i=1}^{n} X_i / E\left\{\sum_{i=1}^{n} X_i\right\}.$$

(a) Find $E(R)$.
(b) Justify an expression for Var(R).
(c) Set up 95 percent confidence limits for R.

(d) How would you use these ideas in analyzing the mortality data for your SMSA? What standard population did you use for purposes of comparison? What second standard population might you have considered?

16. Assuming a uniform distribution of deaths, express q_x^z and m_x^z in terms of population data and deaths. How are q_x^z and m_x^z related? Does it seem reasonable that one is smaller than the other? Why?

17. Explain how a chi-squared goodness of fit test can be used in a mortality study.

2.2.2. Life Table Functions and Distributions

18. Fill out the several columns of the life table below for Bulgaria males, 1965:

x	$_5q_x$	l_x	$_5d_x$	$_5L_x$	$_5m_x$	T_x	$\mathring{e}_x$
20	0.006338	94,864				5,024,927	
25	0.006650						
30	0.008087						
35		92,879					

employing the usual definitions plus: $_5L_x = \frac{5}{2}(l_x + l_{x+5})$; $T_{x+5} = T_x - {}_5L_x$.

19. If the average annual probability of dying between exact ages 20 and 30 is 0.001, what is l_{30}/l_{20}?

20. Prove and interpret

$$\mathring{e}_x = \frac{{}_5L_x}{l_x} + \frac{l_{x+5}}{l_x}\mathring{e}_{x+5}.$$

21. Define in symbols $\bar{x}_D$, the mean age at death in the stationary population. Prove that it is the same as the expectation of life $\mathring{e}_0$, defined as

$$\int_0^\infty l(t)\,dt/l_0.$$

22. Find a usable expression for $_5L_x$ by expanding $l(x + t)$ through Newton's advancing difference formula, and then integrating. Assume $\Delta^3 l_x = \Delta^3 l_{x-5}$.
Hint: Consult Kellison (1975) for Newton's formula.

23. Express d_x as an integral and show that this integral is equivalent to $l_x - l_{x+1}$.

24. The de Moivre model was an early attempt to describe $l_x, 0 \le x \le \omega$. It said that $l_x = a + bx$, $0 \le x \le \omega$, for some values of a and b. What are the values of $_5d_{20}$ and $_5d_{70}$ in such a table?

25. The uniform distribution of deaths (UDD) over 1 year of age assumption says $_tq_x = tq_x, 0 \le t \le 1$. With this assumption, provide a proof (or even two!) that

$$\mu_{x+t} = \frac{q_x}{1 - t \cdot q_x}, \qquad 0 \le t \le 1.$$

26. Show that with UDD,

$$l_{x+t} = l_x - t \cdot d_x, \qquad 0 \le t \le 1.$$

27. Show that with UDD,

$$_{1-s}q_{x+t} = \frac{(1 - s)q_x}{1 - tq_x} = (1 - s)\mu_{x+t}, \qquad 0 \le t \le s \le 1.$$

28. The Balducci distribution of deaths (BDD) was developed by the Italian actuary Gaetano Balducci in 1920. It says that $_{1-t}q_{x+t} = (1 - t)q_x$, $0 \le t \le 1$. Prove that with this assumption,

$$\mu_{x+t} = \frac{q_x}{1 - (1 - t)q_x}, \qquad 0 \le t \le 1.$$

29. Under BDD, show

$$l_{x+t} = \frac{l_x \cdot l_{x+1}}{l_{x+1} + t \cdot d_x}, \qquad 0 \le t \le 1.$$

30. Under BDD, show

$$_tq_x = \frac{t(q_x)}{1 - (1 - t)q_x}, \qquad 0 \le t \le 1.$$

31. Assume a constant force of mortality over a given unit age interval, i.e., $\mu_{x+t} = \mu, 0 \le t \le 1$.
(a) Prove that $_tq_x = 1 - e^{-\mu t}, 0 \le t \le 1$.
(b) Prove that $_{1-t}q_{x+t} = 1 - e^{-\mu(1-t)}, 0 \le t \le 1$.
(c) Prove that $l_{x+t} = l_x \cdot e^{-\mu t}, 0 \le t \le 1$.
(d) Use the results of (a), (b), and (c) to show that $_{1/2}q_x = {}_{1/2}q_{x+1/2} = {}_{1/2}q_{x+1/4}$.

32. (a) Under the constant force assumption, prove that $p_x = (_tp_x)^{1/t}$, $0 < t \le 1$.
(b) Given $p_x = 0.64$, find $_{1/2}p_x$ and $_{1/6}p_x$.

33. Assume that the observed ratios $_nr_x^i = {}_nD_x^i/{}_nD_x$ of dying from cause i also hold in a life table. Let $_np_x^{(-i)}$ be the probability of living from age x to age $x + n$ in a life table derived by eliminating deaths from cause i. Prove that $_np_x^{(-i)} = (_np_x)^{(1-{}_nr_x^i)}$.

34. Refer to Problem 33 above. What further steps are needed to obtain an expression for $\overset{\circ}{e}_x^{(-i)} - \overset{\circ}{e}_x$, the gain in life expectation effected by eliminating cause i.

35. Assuming that deaths are uniformly distributed throughout the year, show that

$$\mathring{e}_x = \tfrac{1}{2}q_x + \tfrac{3}{2}({}_{1}|q_x) + \tfrac{5}{2}({}_{2}|q_x) + \cdots,$$

where ${}_{n}|q_x$ is the probability that a person now aged x will die between ages $x + n$ and $x + n + 1$.

36. Assume $l(x)$ is linear between ages x and $x + n$. Prove that

$$_nq_x = \frac{2n(_nm_x)}{2 + n(_nm_x)}.$$

37. Assume that only one census is available, and it is felt that the population is growing uniformly at rate r. How would you construct a life table?

38. Prove that

$$p_x^{(-i)} = \frac{l_{x+1} + \tfrac{1}{2}(d_x^i)}{l_x - \tfrac{1}{2}(d_x^i)}.$$

39. Let $_n g_x$ equal average amount of time lost between ages x and $x + n$ by those who die during that age interval. Geometrically illustrate why

$$_n g_x = \frac{n(l_x) - {}_nL_x}{{}_nd_x}.$$

40. Assume that $_n\hat{m}_x^A = \frac{9}{1000}$ and $_n\hat{m}_x^B = \frac{14}{1100}$. What statistical hypotheses H_0 and H_1 would you set up? Do you accept H_0 at the 5 percent significance level?

41. The function $\mathring{e}_x$ can be defined in terms of $\sum_{n=0}^{\omega} L_{x+n}$, and l_x. However, it is possible to proceed directly from q_x to $\mathring{e}_x$ by a recursion relation. What is it? Justify its validity. Define the functions which are involved.

42. From the information given below, calculate the probability that a person aged 55 will die before age 60.

			Deaths
Attained age	Exposure	Deaths	Exposure
50–54	20,954	250	0.01193
55–59	15,990	292	0.01826
60–64	10,736	301	0.02804

43. If $l(y)$ is a quadratic, $x - 5 \le y \le x + 5$, express $\mu(x)$ in terms of l_{x-5}, l_x, and l_{x+5}.

44. How fast does $\mu(x)$ have to rise with age for $l(x)$ to be concave downward?

45. Defining $\int_0^n \mu(x + t)\, dt$ as $-\ln(l_{x+n}/l_x)$, show that

$$\int_0^n \mu(x + t)\, dt = q + \frac{q^2}{2} + \frac{q^3}{3} + \cdots$$

without approximation, where $q = 1 - l_{x+n}/l_x$. Show also that

$$_nm_x = \frac{1}{n}\left(q + \frac{q^2}{2} + \frac{q^3}{4} + \cdots\right)$$

wherever the mortality curve is rising with age. Hence show that $_nm_x < \bar{\mu}$, i.e., the life table death rate is less than the mean force of mortality, $e^{-_nn m_x} > e^{-n\bar{\mu}}$, and therefore, to put l_{x+n}/l_x equal to $e^{-_nn m_x}$ unduly raises the curve of l_x over all ages after about 10.

46. Prove the inequalities

(a) $q_x < m_x < \dfrac{q_x}{1 - q_x}$.

(b) $\dfrac{m_x}{1 + m_x} < q_x < m_x$.

47. Prove by integration by parts in $_nL_x = \int_0^n l(x + t)\, dt$ that $_nL_x = nl_{x+n} + {_na_x}(l_x - l_{x+n})$, where $_na_x$ is the mean years lived after age x by those dying before age $x + n$. Justify the same expression in words.

48. If

$$a_x = \frac{L_x - l_{x+1}}{l_x - l_{x+1}} \quad \text{and} \quad m_x = \frac{l_x - l_{x+1}}{L_x}$$

are both known, show how they together determine l_{x+1}/l_x and L_x, and hence that a_x and m_x together suffice to provide all columns of the life table.

49. If the mean age of men dying between 40 and 44 at last birthday is 42.6863, and the age-specific death rate for men 40–44 is 0.004314, find the probability that a man aged 40 will survive to exact age 45. (Data for Austrian males, 1950–1952.)

2.2.3. Modifications to Life Tables

50. A life table is modified by adding 0.001 to $\mu(a)$ at every age. What does this do to the l_x, $_5q_x$, $_5L_x$, and $\mathring{e}_x$?

51. A life table is modified by multiplying $\mu(a)$ for each age by 1.001. What does this do to the several columns?

52. The probability of surviving from birth to age 25 in a particular human population is 0.89. How is this altered if the force of mortality at each age is lowered by 0.0001?

53. From a model mortality table a second table is prepared by multiplying the previous force of mortality by k, a positive integer ≥ 2. Express the new probability $_n\hat{q}_x$ in terms of the old $_nq_x$.

54. How is the probability of living from birth to age 50 affected by a drop in mortality of 0.0001 at age 35? By the same drop at age 55?

55. What difference does it make to the value of $\overset{\circ}{e}_0$ if the forces of mortality between ages 30 and 34 at last birthday drop by 0.001? Assume no changes in $l(x)$, $x \geq 35$.

56. How much does an error of 0.00001 in d_0, total deaths under 1 year of age, ($l_0 = 1$) affect $\overset{\circ}{e}_0$?

57. If infant mortality was reduced by 20 per thousand births in a population, and all other mortality continued unchanged at $\overset{\circ}{e}_0 = 70$, by how much would $\overset{\circ}{e}_0$ rise?

2.2.4. Force of Mortality and Expectation of Life

58. If the force of mortality $\mu(x)$ is age-independent and equal to 0.02, in how many years will an initial group of 100,000 babies be reduced to 80,000?

59. A man aged 30 marries a woman aged 25. Express in symbols the expected number of years they will both be alive. If μ_x is constant and equal to 0.02, evaluate the symbolic expression.

60. (a) Prove that if at age x and above, mortality is fixed at μ, then the expectation of life at age x is $1/\mu$.

(b) Hence that a decrease by the factor $1 - \delta$, $\delta > 0$, in μ raises the expectation of life to $1/\mu(1 - \delta)$; i.e., if δ is small subtracting $\delta\mu$ from μ raises $\overset{\circ}{e}_x$ by approximately $\overset{\circ}{e}_x\delta$. In numbers, a 1 percent decrease in deaths raises the expectation of life by 1 percent.

(c) If $\mu(a)$ is rising for $a > x$, what is an upper bound on $\overset{\circ}{e}_x$? How is this changed if $\mu(a)$ is replaced by $(1 - \delta)\mu(a)$, $a > x$?

61. Suppose that within an age interval $(x, x + n)$ the force of mortality rises in the parabolic form

$$\mu(x + t) = {_nM_x} + 0.024t^2 {_nM_x^2}.$$

Show that this produces exactly the life table due to Reed and Merrell (1939), in which

$$_nq_x = 1 - \exp[-(n_nM_x + 0.008n^3 {_nM_x^2})].$$

62. The assumption that M_x is the constant value of $\mu(x + t)$ at each age within the interval $(x, x + 1)$ translates into a probability of living $l_{x+1}/l_x = e^{-M_x}$. The assumption that $_{1-t}q_{x+t} = (1 - t)q_x$, $0 < t < 1$ (known as the Balducci assumption) is equivalent to l_x, l_{x+t}, and l_{x+1} being related as $1/l_{x+t} = (t/l_{x+1}) + (1 - t)/l_x$. Prove both of these statements. What do they respectively imply about the shape of $l(x + t)$ in the interval $0 \leq t \leq 1$?

63. The force of mortality μ_x in a certain population increases linearly by 0.002 per year from age 40, starting at $\mu_{40} = 0.01$. What is the probability that a man aged 40 will live to age 70?

64. Assuming a constant force of mortality, k, derive an expression for the error involved in the two approximate relations below, and give numerical values of the errors when $k = 0.01$.

(i)
$$q_x = \frac{2\mu_{x+1/2}}{2 + \mu_{x+1/2}},$$

(ii)
$$\mu_x = \frac{d_{x-1} + d_x}{2l_x}.$$

Show that if the law followed by the force of mortality were $\mu_x = 1/(A - x)$, the foregoing relations would hold exactly (A is a constant). (Society of Actuaries.)

65. In a stationary population subject to

$$l(x) = \frac{e^{-0.02x} + 1 - 0.01x}{2}, \qquad 0 \le x \le \omega$$

what is the value of $\overset{\circ}{e}_0$? What is the mean age of death? The mean age of the living? At what age is the absolute number of deaths a maximum? At what age does the last person die?

66. Suppose that $l_6 = 0.64$, $l_{16} = 0.32$, $l_{26} = 0.16, \ldots, l_{66} = 0.01$. With the probability of dying over each 10-year interval equal to $\frac{1}{2}$, calculate a suitable interpolation to age x, where $6 < x < 66$. Find the chance that of three persons aged 6 at least two will be alive 20 years later.

67. The values l_x of a life table in which no one lives beyond age 100 are given by

$$l_0 = 1; l_5 = (\tfrac{1}{2})^{0.1}; l_{10} = (\tfrac{1}{3})^{0.1}; \ldots;$$

$$l_{5i} = \left(\frac{1}{i + 1}\right)^{0.1}; \ldots; l_{100} = 0.$$

Find $\overset{\circ}{e}_{60}$, assuming a straight line between the given values of l_x, and check your answer by integration of the curve

$$l(x) = \left(\frac{1}{x/5 + 1}\right)^{0.1}.$$

What is $\overset{\circ}{e}_0$ on this life table? What is $\mu(x)$? What is $_5q_x$?

68. On the life table with $l(x) = (100 - x)/190$, $5 \le x \le 100$, work out
(a) the chance that a child who has reached age 5 will live to 60,
(b) the chance that a man of 30 lives to age 80,
(c) the probability of dying within five years for a man aged 40, i.e., $_5q_{40}$,
(d) the average age at death of those dying between ages 40 and 45,

(e) the instantaneous death rate at age 40, i.e., μ_{40},

(f) the expectation of life at age 40,

(g) the chance that of three men aged 30 at least one survives to age 80.

69. If $l(x) = \sqrt{1 - x/100}$, what is: (a) $\mu(x)$; (b) $\overset{\circ}{e}_0$?

70. Prove that

$$\int_0^{\omega - x} l(x + t)\mu(x + t)\, dt = l_x,$$

$$\int_0^{\omega - x} t l(x + t)\mu(x + t)\, dt = \int_0^{\omega - x} l(x + t)\, dt = l_x \overset{\circ}{e}_x.$$

71. The following table contains four typographical errors, each of a single digit. What are they?

	l_x	$_5 d_x$	$_5 q_x$	$_5 L_x$	T_x	$\overset{\circ}{e}_x$
20	95,722	857	0.008953	476,495	4,922,814	51.533
25	94,265	699	0.007368	472,566	4,456,319	47.975
30	94,166	800	0.008496	468,895	3,983,754	42.306

2.2.5. Laws of Gompertz, Makeham, and Weibull

72. A mortality table that follows Gompertz's Law is found to have inadequate mortality rates. It is proposed to construct a new table following Makeham's Law with the same values of B and c. Find an approximation for the difference between the old and the new mortality probabilities in terms of the force of mortality and the constant A. (Society of Actuaries.)

73. Show that the Weibull distribution of death rates

$$\mu(a) = \frac{\beta}{\theta}\left(\frac{a}{\theta}\right)^{\beta - 1}$$

is equivalent to the survival function

$$l(a) = \exp\left[-\left(\frac{a}{\theta}\right)^{\beta}\right].$$

Given the age-specific rates for United States males, 1967:

	x	$_5 M_x$
30		0.00220, say equal to $\mu(32.5)$
60		0.02805, say equal to $\mu(62.5)$

find β, θ, and hence $l(90)$.

74. Assuming a mortality table under which the force of mortality consists of three elements—one a constant, another increasing in arithmetical progression throughout life, and the third increasing in geometrical progression throughout life, find expressions for l_x and l_{x+n}/l_x. (Society of Actuaries.)

75. If $\mu(x) = A + Bc^x$, find an expression for $d\mathring{e}_x/dA$ which does not contain A, B, or c. (Society of Actuaries.)

2.3. Solutions

1. For Indiana, $46,750/6,000,000 = 0.0078$.
   For Muncie, $567/71,500 = 0.0079$.

2. For Indiana,

$$_1m_0 = 0.0075, \quad _4m_1 = 0.0075, \quad _{15}m_5 = 0.0080, \quad _{80}m_{20} = 0.0078;$$
$$_1m_0^m = 0.0080, \quad _4m_1^m = 0.0080, \quad _{15}m_5^m = 0.0087, \quad _{80}m_{20}^m = 0.0080;$$
$$_1m_0^f = 0.0070, \quad _4m_1^f = 0.0070, \quad _{15}m_5^f = 0.0073, \quad _{80}m_{20}^f = 0.0076.$$

For Muncie,

$$_1m_0 = 0.0080, \quad _4m_1 = 0.0075, \quad _{15}m_5 = 0.0080, \quad _{80}m_{20} = 0.0080;$$
$$_1m_0^m = 0.0080, \quad _4m_1^m = 0.0080, \quad _{15}m_5^m = 0.0085, \quad _{80}m_{20}^m = 0.0118;$$
$$_1m_0^f = 0.0080, \quad _4m_1^f = 0.0070, \quad _{15}m_5^f = 0.0075, \quad _{80}m_{20}^f = 0.0056.$$

3.

$$\frac{\sum\limits_x P_x^s \cdot m_x}{\sum\limits_x P_x^s} = \frac{800 + 3000 + 12,000 + 32,000}{6,000,000} = 0.007967.$$

4.

$$\frac{D}{\sum\limits_x {}_nm_x^s \cdot {}_nP_x}(0.0078) = \frac{567}{559.45}(0.0078) = 0.0079.$$

5.

$$\frac{\sum\limits_x ({}^mP_x^s \cdot {}^mm_x + {}^fP_x^s \cdot {}^fm_x)}{6,000,000} = 0.0079.$$

6.

$$\mathrm{Var}(M) = \sum_x \left(\frac{P_x}{P}\right)^2 \mathrm{Var}(M_x)$$

$$= \sum_x \left(\frac{P_x}{P}\right)^2 \mathrm{Var}\frac{\sum\limits_{i=1}^{P_x} X_i}{P_x},$$

where $\{X_i\}$ are independent, each with distribution $P(X_i = 1) = M_x$,

$P(X_i = 0) = 1 - M_x$

$$= \sum_x \left(\frac{P_x}{P}\right)^2 \frac{M_x(1 - M_x)}{P_x}$$

$$= \sum_x \frac{D_x(1 - M_x)}{P^2}.$$

16. $q_x^z = D_x^z/[0.5(P_x^z + P_x^{z+1} + D_x^z)]$,
$m_x^z = D_x^z/P_x^{z+1/2}$,
$q_x^z = m_x^z/(1 + \frac{1}{2}m_x^z)$. Yes, the exposed to risk in q_x^z is larger than in m_x^z.

18.

x	$_5q_x$	l_x	$_5d_x$	$_5L_x$	$_5m_x$
20	0.006338	94,864	601	472,817.5	0.001271
25	0.006650	94,263	627	469,747.5	0.001335
30	0.008087	93,636	757	466,287.5	0.001623
35		92,879			

x	T_x	$\overset{\circ}{e}_x$
20	5,024,927	52.97
25	4,552,109.5	48.29
30	4,082,362	43.60
35	3,616,074.5	38.93

19. 0.99.

20.

$$\overset{\circ}{e}_x = \frac{T_x}{l_x} = \frac{_5L_x + T_{x+5}}{l_x} = \frac{_5L_x}{l_x} + \frac{l_{x+5}}{l_x}\overset{\circ}{e}_{x+5}.$$

The complete expectation of life at age x is the average years lived between x and $x + 5$ plus the complete expectation of life beyond age $x + 5$ to those who live to $x + 5$.

21.

$$\bar{x}_D = \int_0^\omega \frac{t l_t \mu_t \, dt}{l_0}.$$

After integration by parts,

$$\bar{x}_D = \int_0^\omega \frac{l_t}{l_0} \, dt = \overset{\circ}{e}_0.$$

22.

$$l_{x+t} = E_5^{t/5} l_x = \left(1 + \underset{5}{\Delta}\right)^{t/5} l_x$$

$$= \left\{ l_x + \frac{t}{5} \underset{5}{\Delta} l_x + \frac{t(t-5)}{2! \, 5^2} \underset{5}{\Delta^2} l_x + \frac{t(t-5)(t-10)}{3! \, 5^3} \underset{5}{\Delta^3} l_{x-5} + \cdots \right\},$$

$$_5L_x = \int_0^5 l(x+t) \, dt$$

$$\doteq 5l_x + \frac{25}{10} \underset{5}{\Delta} l_x - \frac{5}{12} \underset{5}{\Delta^2} l_x + \frac{5}{24} \underset{5}{\Delta^3} l_{x-5}$$

$$= \frac{-5}{24} l_{x-5} + \frac{65}{24} l_x + \frac{65}{24} l_{x+5} - \frac{5}{24} l_{x+10}.$$

23.

$$d_x = \int_0^1 l(x+t)\mu(x+t) \, dt$$

$$= \int_0^1 (-1) dl(x+t)$$

$$= l_x - l_{x+1}.$$

after integration by parts.

24. Both equal $-5b$. (Clearly, $b < 0$.)

25. Proof 1.

$$_tp_x = 1 - t(q_x),$$

$$\frac{d}{dt}(_tp_x) = -q_x = -_tp_x\mu_{x+t},$$

$$\mu_{x+t} = \frac{q_x}{_tp_x} = \frac{q_x}{1 - tq_x}.$$

Proof 2.

$$_tq_x = 1 - \exp\left(-\int_0^t \mu_{x+z} \, dz\right),$$

$$-\int_0^t \mu_{x+z} \, dz = \ln_e(1 - tq_x),$$

$$\mu_{x+t} = \frac{q_x}{1 - tq_x}.$$

27.

$$1-s q_{x+t} = \frac{l_{x+t} - l_{x+t+1-s}}{l_{x+t}}$$

$$= \frac{l_x - td_x - l_x + (1 - s + t)d_x}{l_x - td_x}$$

$$= \frac{(1 - s)q_x}{1 - tq_x}$$

$$= (1 - s)\mu_{x+t}.$$

28.

$$1-t q_{x+t} = 1 - \exp\left(-\int_t^1 \mu_{x+z}\, dz\right),$$

$$-\int_t^1 \mu_{x+z}\, dz = \ln_e[1 - (1 - t)q_x].$$

Differentiate both sides w/r/t t.

$$\mu_{x+t} = \frac{q_x}{1 - (1 - t)q_x}.$$

31. (a) $_tq_x = 1 - \exp\left(-\int_0^t \mu_{x+z}\, dz\right)$

$$= 1 - e^{-\mu t}.$$

32. (a) $p_x = e^{-\mu} = (e^{-\mu t})^{1/t} = (_tp_x)^{1/t}.$
(b) 0.8 and 0.9283.

33. Let μ_y^i = force of mortality from cause i. Assume $\mu_y^i/\mu_y = A$ for $x < y < x + n.$

$$_np_x = \exp\left\{-\int_0^n \mu_{x+t}\, dt\right\} \quad \text{and} \quad \log{_np_x} = -\int_0^n \mu_{x+t}\, dt.$$

Likewise,

$$\log {_np_x^{(-i)}} = -\int_0^n \mu_{x+t}^{(-i)}\, dt.$$

By definition,

$$\int_0^n \mu_{x+t}^{(-i)}\, dt = \int_0^n \{\mu_{x+t} - \mu_{x+t}^i\}\, dt$$

$$= (1 - A)\int_0^n \mu_{x+t}\, dt$$

$$= (1 - A)(-1)\log {_np_x}.$$

So

$$\log {}_np_x^{(-i)} = (1 - A) \log {}_np_x.$$

To show that $A = {}_nr_x^i$,

$$_nr_x^i = \frac{{}_nd_x^i}{{}_nd_x} = \frac{\int_0^n l_{x+t}\mu_{x+t}^i \, dt}{\int_0^n l_{x+t}\mu_{x+t} \, dt} = A.$$

Thus

$$\log {}_np_x^{(-i)} = \log({}_np_x)^{(1 - {}_nr_x^i)} \quad \text{and} \quad {}_np_x^{(-i)} = {}_np_x^{(1 - {}_nr_x^i)}.$$

37.

$$p_{x+1/2} = \frac{P_{x+1}^z}{P_x^{z-1}} = \frac{P_{x+1}^z}{P_x^z/(1 + r)}.$$

Interpolate to get p's at integral ages. After choosing l_0, $l_x = l_0 p_x$, $d_x = l_x - l_{x+1}$, $q_x = 1 - p_x$ for $x = 1, 2, \ldots$.

38. Those who do not die from the ith cause would be subject to all other causes for half a year on the average:

$$q_x^{(-i)} = \frac{d_x - d_x^i + (\frac{1}{2}d_x^i)q_x^{(-i)}}{l_x}.$$

Thus

$$q_x^{(-i)} = \frac{d_x - d_x^i}{l_x - \frac{1}{2}d_x^i} \quad \text{and} \quad p_x^{(-i)} = 1 - q_x^{(-i)} = \frac{l_{x+1} + \frac{1}{2}(d_x^i)}{l_x - \frac{1}{2}(d_x^i)}.$$

39.

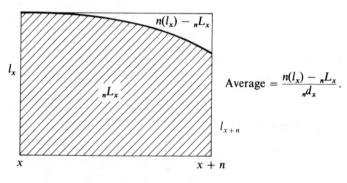

$$\text{Average} = \frac{n(l_x) - {}_nL_x}{{}_nd_x}.$$

41. Start at the highest age where $q_x < 1$.

$$\mathring{e}_x = 1 - q_x + f_x,$$

where f_x is the expected fraction of the year of age from x to $x + 1$ of those who die before their $(x + 1)$st birthday.

Move backwards with age.

$$\mathring{e}_x = (1 - q_x)(\mathring{e}_{x+1} + 1) + f_x q_x.$$

This step is justified by the addition law for probabilities.

42. Assume that the observed rates $_5M_x$ equal the life table rates $_5m_x$. Also assume that l_x is linear in each 5-year age group. By Problem 36, $_5q_{55} = 0.08731$, $_5q_{50} = 0.05792$, and $_5q_{60} = 0.13102$. The last two values attest to the reasonableness of $_5q_{55}$'s value.

43. Assume that $l(y) = A + By + Cy^2$, for $x - 5 \le y \le x + 5$. Then

$$\mu(y) = -\frac{1}{l(y)}\frac{dl(y)}{dy} = -\frac{B + 2Cy}{l(y)}.$$

By manipulating $l(x - 5)$, $l(x)$, and $l(x + 5)$ one can obtain

$$2C = \frac{l(x + 5) - 2l(x) + l(x - 5)}{25},$$

and

$$B = \frac{l(x) - l(x - 5)}{5} - (2x - 5)\left[\frac{l(x + 5) - 2l(x) + l(x - 5)}{50}\right],$$

$$\mu_y = \frac{1}{l(y)}\left[\frac{l(x - 5) - l(x)}{5} + \frac{l(x + 5) - 2l(x) + l(x - 5)}{50}\{2x - 2y - 5\}\right].$$

When $y = x$, the above becomes $\mu_x = (1/l_x)[0.1l_{x-5} - 0.1l_{x+5}]$.

44. One needs $d^2l(x)/dx^2 < 0$ for $l(x)$ to be concave downward. Since

$$\frac{d^2l(x)}{dx^2} = l(x)\left[\mu^2(x) - \frac{d\mu(x)}{dx}\right],$$

this is equivalent to requiring $d\mu(x)/dx > \mu^2(x)$.

45.

$$\int_0^n \mu(x + t)\,dt = -\ln\left(\frac{l_{x+n}}{l_x}\right)$$

$$= -\ln(1 - q)$$

$$= q + \frac{q^2}{2} + \frac{q^3}{3} + \cdots.$$

If $l(x)$ is linear between ages x and $x + n$,

$$_nq_x = \frac{n(_nm_x)}{1 + (n/2)(_nm_x)}.$$

Solve this for $n(_nm_x)$ to obtain $_nq_x/[1 - \frac{1}{2}(_nq_x)]$. The denominator can be expressed as a series, and

$$_nm_x = \frac{1}{n}\left(q + \frac{q^2}{2} + \frac{q^3}{4} + \cdots\right).$$

Therefore,

$$_nm_x < \frac{1}{n} \int_0^n \mu(x + t)\, dt = \bar{\mu},$$

and

$$\exp\{-n(_nm_x)\} > \exp\{-n\bar{\mu}\}.$$

Hence,

$$\exp\{-n(_nm_x)\} > \frac{l_{x+n}}{l_x}.$$

46. (a) With rare exceptions,

$$l_x > \int_0^1 l(x + t)\, dt > l_{x+1}.$$

Hence,

$$\frac{d_x}{l_x} < \frac{d_x}{\int_0^1 l(x + t)\, dt} < \frac{d_x}{l_{x+1}},$$

or

$$q_x < m_x < \frac{q_x}{1 - q_x}.$$

(b) The inequality $m_x < q_x/(1 - q_x)$, and routine algebra yield $m_x/(1 + m_x) < q_x$. The second part was proved in part (a).

47. Let $U = l(x + t)$ and $dV = dt$ in $_nL_x$. Then

$$\int_0^n l(x + t)\, dt = tl(x + t)\Big|_0^n - \int_0^n t\, dl(x + t)$$

$$= nl_{x+n} + {}_na_x(l_x - l_{x+n}),$$

where

$$_na_x = \frac{\int_0^n tl(x + t)\mu(x + t)\, dt}{l_x - l_{x+n}}.$$

The aggregate number of years lived between ages x and $x + n$ by the l_x starters is $_nL_x$. This aggregate number is also equal to n times the l_{x+n} enders plus $_na_x$ times those who died between ages x and $x + n$.

48. $L_x = (l_x - l_{x+1})/m_x$. Hence,

$$a_x(l_x - l_{x+1}) = \frac{l_x - l_{x+1}}{m_x} - l_{x+1},$$

or

$$l_{x+1} = \frac{l_x(1 - m_x a_x)}{1 + m_x(1 - a_x)}.$$

49. From Solution 47 and 48,

$$l_{x+n} = \frac{l_x[1 - {}_nm_x \cdot {}_na_x]}{1 + n({}_nm_x) - {}_nm_x({}_na_x)}.$$

With the given data,

$$_5p_{40} = \frac{0.9884}{1.0100} = 0.9786.$$

50. Let $\hat{l}_x$, $_5\hat{q}_x$, $_5\hat{L}_x$, and $\overset{\circ}{\hat{e}}_x$ be the modified functions. Then

$$\hat{l}_x = e^{-0.001x}l_x,$$

$$_5\hat{q}_x = (1 - e^{-0.005}) + e^{-0.005}\,{}_5q_x$$
$$\doteq 0.005 + 0.995({}_5q_x),$$

$$_5\hat{L}_x = \int_0^5 e^{-0.001(x+t)}l(x+t)\,dt,$$

and

$$\overset{\circ}{\hat{e}}_x = \frac{1}{l_x}\int_0^\infty e^{-0.001t}l(x+t)\,dt.$$

51.

$$\hat{l}_x = l_0 e^{-1.001\int_0^x \mu_y\,dy} \quad (= (l_x)^{1.001} \text{ if } l_0 = 1),$$

$$_5\hat{q}_x = 1 - {}_5\hat{p}_x = 1 - ({}_5p_x)^{1.001},$$

$$_5\hat{L}_x = \hat{l}_x \int_0^5 e^{-1.001\int_x^{x+t}\mu_y\,dy}\,dt,$$

$$\overset{\circ}{\hat{e}}_x = \int_0^\infty e^{-1.001\int_x^{x+t}\mu_y\,dy}\,dt.$$

52.

$$_{25}p_0 = \exp\left\{-\int_0^{25}\mu(t)\,dt\right\} = 0.89,$$

$$_{25}\hat{p}_0 = \exp\left\{-\int_0^{25}[\mu(t) - 0.0001]\,dt\right\}$$
$$= e^{0.0025}(0.89) = 0.89223.$$

53.

$$_n\hat{q}_x = 1 - \exp\left\{-k\int_0^n\mu(x+t)\,dt\right\}$$
$$= 1 - [_np_x]^k$$
$$= 1 - [1 - {}_nq_x]^k$$
$$\doteq k({}_nq_x) - \frac{k(k-1)}{2}({}_nq_x)^2.$$

54. $_{50}p_0 = p_0 p_1 p_2 \cdots p_{34} p_{35} p_{36} \cdots p_{49}$. New $\hat{q}_{35} = q_{35} - 0.0001$, or $\hat{p}_{35} = p_{35} + 0.0001$. Hence

$$_{50}\hat{p}_0 = {}_{50}p_0 + 0.0001 \, \frac{l_{35}}{l_0} \cdot \frac{l_{50}}{l_{36}}.$$

The same drop at age 55 would have no effect on $_{50}p_0$.

55.

$$\mathring{e}_0 = \int_0^\omega l(t) \, dt / l(0)$$

$$= \left\{ \int_0^{30} l(t) \, dt + \int_{30}^{35} l(t) \, dt + \int_{35}^\omega l(t) \, dt \right\} \bigg/ l(0).$$

With the change in forces of mortality,

$$\int_{30}^{35} l(t) \, dt = \int_{30}^{35} l(0) \exp\left\{ -\int_0^t [\mu(y) - 0.001] \, dy \right\} dt$$

$$= \int_{30}^{35} l(0) e^{0.001t} \exp\left\{ -\int_0^t \mu(y) \, dy \right\} dt$$

$$\doteqdot e^{0.001(32.5)} \int_{30}^{35} l(0) \exp\left\{ -\int_0^t \mu(y) \, dy \right\} dt$$

by an approximation suggested by a Mean Value Theorem for Integrals

$$\doteqdot 1.033 \int_{30}^{35} l(t) \, dt.$$

Thus the new expectation of life equals the old value, increased by

$$0.033(_5L_{30})/l(0).$$

56. $\mathring{e}_0 \doteqdot e_0 + \frac{1}{2}$.
$\quad e_0 = p_0 + p_0 e_1 = (1 - d_0)(1 + e_1)$ since $l_0 = 1$. Therefore, an error of 0.00001 in d_0 affects $\mathring{e}_0$ by $\pm 0.00001(1 + e_1)$.

57. Under the old conditions, $e_0 = p_0(1 + e_1) = 69.5$, or $1 + e_1 = 69.5/p_0$. The old probability $q_0 = 0.020 + x$ for some x. The old and new e_0 values are

$$e_0 = (1 - 0.020 - x)(1 + e_1),$$

and

$$e_0 = (1 - x)(1 + e_1).$$

Thus $\mathring{e}_0$ rises by $0.020[(69.5)/p_0]$.

58. $80,000 = 100,000e^{-0.02n} \Rightarrow n \doteq 11.$

59.

$$\mathring{e}_{30:\overline{25}|} = \int_0^\infty {}_tp_{30:\overline{25}|} \, dt$$

$$= \int_0^\infty {}_tp_{30} \cdot {}_tp_{25} \, dt.$$

If μ_x is constant and equal to 0.02,

$$\mathring{e}_{30:\overline{25}|} = \int_0^\infty e^{-0.02t} e^{-0.02t} \, dt = 25 \text{ years.}$$

60. (a)

$$\mathring{e}_x = \frac{1}{l_x} \int_0^\infty l(x+t) \, dt$$

$$= \int_0^\infty \exp\left\{-\int_0^t \mu(x+w) \, dw\right\} dt$$

$$= \int_0^\infty e^{-\mu t} \, dt = \frac{1}{\mu}.$$

(b) The new expectation of life $= 1/\mu(1 - \delta)$. Since $1/(1 - \delta) \doteq 1 + \delta$, the new expectation $\mathring{e}_x \doteq (1 + \delta)\mathring{e}_x$.

(c)

$$\mathring{e}_x = \int_0^\infty \exp\left\{-\int_0^t \mu(x+w) \, dw\right\} dt$$

$$< \int_0^\infty \exp\{-t\mu(x)\} \, dt$$

$$= \frac{1}{\mu(x)}.$$

Replacing $\mu(x + w)$ by $(1 - \delta)\mu(x + w)$ for $w \geq 0$ yields a new upper bound of $1/(1 - \delta)\mu(x)$, or approximately $(1 + \delta)[1/\mu(x)]$ for small δ.

61.

$${}_nq_x = 1 - {}_np_x$$

$$= 1 - \exp\left\{-\int_0^n \mu(x+t) \, dt\right\}$$

$$= 1 - \exp\{-n({}_nM_x) - 0.008n^3({}_nM_x^2)\}.$$

62.

$$\frac{l_{x+1}}{l_x} = \exp\left\{-\int_0^1 \mu(x+t)\,dt\right\}$$

$$= \exp\{-M_x\}.$$

$$_{1-t}q_{x+t} = \frac{l(x+t) - l(x+1)}{l(x+t)}.$$

When

$$_{1-t}q_{x+t} = (1-t)q_x,$$

then

$$1 - \frac{l(x+1)}{l(x+t)} = (1-t) - (1-t)\frac{l(x+1)}{l(x)},$$

or

$$\frac{1}{l(x+t)} = \frac{t}{l(x+1)} + \frac{1-t}{l(x)}.$$

The constant force assumption yields an exponential curve.

$$l(x+t) = l(x)\exp\{-tM_x\}, \qquad 0 \le t \le 1.$$

The Balducci assumption yields part of a hyperbola for $l(x+t)$:

$$l(x+t) = \frac{l(x)\cdot l(x+1)}{l(x+1) + t\,d(x)}, \qquad 0 \le t \le 1.$$

63.

$$_{30}p_{40} = \exp\left(-\int_0^{30} \mu_{40+t}\,dt\right)$$

$$= \exp\left\{-\int_0^{30}(0.01 + 0.002t)\,dt\right\}$$

$$\doteq 0.3012.$$

64. By Problem 31, $q_x = 1 - e^{-k}$. For part (i), the error equals

$$1 - e^{-k} - \frac{k}{1 + 0.5k},$$

and when $k = 0.01$, the numerical error is $-0.000\,000\,08$. For part (ii), the approximate expression equals

$$\frac{1}{2}\left\{q_{x-1}\frac{l_{x-1}}{l_x} + q_x\right\} = \frac{1}{2}\{(1-e^{-k})e^k + (1-e^{-k})\}$$

$$= \frac{1}{2}(e^k - e^{-k}).$$

The error $= k - \frac{1}{2}(e^k - e^{-k})$, and when $k = 0.01$, the numerical error is $-0.000\,000\,17$.

If $\mu_x = 1/(A - x)$, then $p_x = 1 - 1/(A - x)$, and $q_x = 1/(A - x)$. Thus,

$$\frac{2\mu_{x+1/2}}{2 + \mu_{x+1/2}} = \frac{1}{A - x} = q_x.$$

$$l_x = l_0 \exp\left\{-\int_0^x \mu(t)\, dt\right\}$$

$$= l_0(1 - x/A).$$

Hence

$$d_x = \frac{l_0}{A} \quad \text{and} \quad d_{x-1} = \frac{l_0}{A}.$$

Therefore,

$$\frac{d_{x-1} + d_x}{2l_x} = \frac{1}{A - x} = \mu_x.$$

65. The last person dies at age 110 (last birthday).
$\mathring{e}_0 = \int_0^{110} l(x)\, dx$, since $l(0) = 1$. Thus $\mathring{e}_0 = 46.98$.

$$\text{The mean age of death} = \frac{\int_0^{110} tl(t)\mu(t)\, dt}{\int_0^{110} l(t)\mu(t)\, dt}$$

$$= \frac{\int_0^{110} t[-dl(t)]}{\int_0^{110} (-1)dl(t)}$$

$$= 46.98$$

$$\text{The mean age of the living} = \frac{\int_0^{110} tl(t)\, dt}{\int_0^{110} l(t)\, dt} = \frac{1613.46}{46.98} = 34.34.$$

The function $d(x) = l(x) - l(x + 1)$ has a derivative of

$$-0.01\{e^{-0.02x} - e^{-0.02(x+1)}\}.$$

Therefore, the number of deaths is maximal at age 0, where $d(0) = 0.0149$.

66.

$$l_x = l_6(\tfrac{1}{2})^{(x-6)/10}, \qquad 6 < x < 66.$$

$$\left(\frac{l_{26}}{l_6}\right)^3 + \binom{3}{2}\left(\frac{l_{26}}{l_6}\right)^2\left(1 - \frac{l_{26}}{l_6}\right) = \frac{5}{32}.$$

67.

$$\mathring{e}_{60} = \frac{1}{l_{60}} \int_{60}^{100} l(x)\, dx$$

$$= \frac{5}{l_{60}}\left[\frac{l_{60} + l_{65}}{2}\right] + \frac{5}{l_{60}}\left[\frac{l_{65} + l_{70}}{2}\right] + \cdots + \frac{5}{l_{60}}\left[\frac{l_{95} + l_{100}}{2}\right]$$

$$= 36.6 \text{ years.}$$

As a check, $\int_{60}^{95} (0.2x + 1)^{-0.1} \, dx = 26.466$ and $5[(l_{95} + l_{100})/2] = 1.853$, and thus $(1/l_{60})[26.466 + 1.853] = 36.6$ years.

$$\mathring{e}_0 = \frac{1}{l_0} \int_0^{95} (0.2x + 1)^{-0.1} \, dx + 1.853 = 78.6 \text{ years},$$

$$\mu_x = 0.02(0.2x + 1)^{-1},$$

$$_5 q_x = 1 - \exp\left\{ -0.02 \int_0^5 \frac{dt}{0.2(x + t) + 1} \right\}$$

$$= 1 - \exp\left\{ -0.1 \ln\left(\frac{0.2x + 2}{0.2x + 1} \right) \right\}$$

$$= 1 - \left(\frac{0.2x + 1}{0.2x + 2} \right)^{0.1}.$$

68. (a) 0.421.   (b) 0.286.   (c) 0.083.

(d) $\displaystyle\int_0^5 \frac{t l(40 + t)\mu(40 + t) \, dt}{l(40) - l(45)} = 2.5$. Hence, the average age at death is 42.5.

(e) 1/60.   (f) $\mathring{e}_{40} = 30$.
(g) P[at least one survives] $= 1 - P$[all die]

$$= 1 - \left[\frac{l_{30} - l_{80}}{l_{30}} \right]^3 \doteq 0.64.$$

69. (a)

$$\mu_x = \frac{1}{200} \cdot \frac{1}{(1 - x/100)}.$$

(b)

$$\mathring{e}_0 = \int_0^{100} \left(1 - \frac{x}{100} \right)^{1/2} dx = 66.67 \text{ years}.$$

70.

$$\int_0^{\omega - x} l(x + t)\mu(x + t) \, dt = -\int_0^{\omega - x} dl(x + t) = l_x.$$

$$\int_0^{\omega - x} t l(x + t)\mu(x + t) \, dt = -t l(x + t) \Big|_0^{\omega - x} + \int_0^{\omega - x} l(x + t) \, dt$$

by integration by parts,

$$= l_x \mathring{e}_x.$$

71. The corrected typographical errors are underlined as follows: $T_{20} = 4932814$, $l_{25} = 94865$, $\mathring{e}_{25} = 46.975$, and $T_{30} = 3,983,753$.

72.

$$q_x^{\text{New}} - q_x^{\text{Old}} = \exp\left\{-\int_0^1 Bc^{x+t}\, dt\right\} - \exp\left\{-\int_0^1 [A + Bc^{x+t}]\, dt\right\}$$

$$= (1 - e^{-A}) \exp\left\{-\int_0^1 \mu_{x+t}^{\text{Old}}\, dt\right\}$$

$$\doteqdot A \exp\left\{-\int_0^1 \mu_{x+t}^{\text{Old}}\, dt\right\}.$$

73. Assume that $l(0) = 1$. Then

$$l(a) = \exp\left\{-\int_0^a \mu(x)\, dx\right\}$$

$$= \exp\left[-\left(\frac{a}{\theta}\right)^\beta\right].$$

The assumptions $\mu(32.5) = (\beta/\theta^\beta)(32.5)^{\beta-1}$ and $\mu(62.5) = (\beta/\theta^\beta)(62.5)^{\beta-1}$ yield $\beta = 4.89269$. Further calculations produce $\theta = 77.08719$. Hence, $l(90) = 0.11843$.

74. Assume that $\mu_x = A + Bx + c^x$. Then

$$l_x = l_0 \exp\left\{-\int_0^x (A + Bt + c^t)\, dt\right\}$$

$$= l_0 \exp\left\{-\left(Ax + \frac{B}{2}x^2 + \frac{c^x - 1}{\ln_e c}\right)\right\}.$$

$$\frac{l_{x+n}}{l_x} = \exp\left\{-\left(An + Bxn + \frac{B}{2}n^2 + \frac{c^x(c^n - 1)}{\ln_e c}\right)\right\}.$$

75. We prove a more general proposition, applying to any additive constant in the force of mortality, which we may call $\mu(x) = A + \mu^*(x)$. The derivative of $l(x + t)/l(x)$ is

$$\frac{dl(x + t)/l(x)}{dA} = \frac{d\,\exp(-\int_0^t (A + \mu^*(x + s))\, ds)}{dA}$$

$$= \frac{d}{dA} e^{-At} \exp\left(-\int_0^t \mu^*(x + s)\, ds\right)$$

$$= -tl(x + t)/l(x).$$

Hence using the fact that the derivative of an integral is the integral of the derivative for our integrand, we have

$$\frac{d\mathring{e}(x)}{dA} = \frac{d \int_0^{\omega-x} l(x+t)/l(x)\, dt}{dA}$$

$$= \int_0^{\omega-x} \frac{dl(x+t)/l(x)\, dt}{dA} = -\int_0^{\omega} \frac{tl(x+t)\, dt}{l(x)}.$$

This does not involve the constants of the Makeham or any other graduation.

CHAPTER III
Uses of Stable Theory

The problems of this chapter require no more apparatus than the stable age distribution expressed in terms of the birth rate b, the rate of increase r, and the probability $l(x)$ of surviving from birth to age x. If these quantities are fixed over time the fraction of the population between age x and $x + dx$ is $be^{-rx}l(x)\,dx$, a proposition due to Euler (1760). The population is supposed closed to migration.

From this expression we can show how any of the rates of birth and death affect the age distribution. For example, the quantity $be^{-rx}l(x)$ or its integral over any range of ages can be differentiated with respect to any of the age-specific rates and the effect of a small change in any rate calculated. As between two countries, or two points of time in the same country, we can permute the births and deaths and find what part of the difference in age distribution is due to births and what part to deaths. The expression can be converted into a form that shows explicitly the moments of the distribution of ages. Doing this permits expressing the fraction of the population in any age interval in terms of the moments of the distribution as a whole and of the part of the distribution in that interval. The sensitivity of the fraction in any age interval to the several moments is thus immediately available.

From this theory one can go on to the inverse problem: inferring the rate of increase and other features from a knowledge of the fraction in an age interval, plus some assumptions regarding the moments.

The identical theory can be applied to a sub-population, for instance the women bearing children in a particular year, the labor-force, children at school. All that is needed is an additional factor, say $m(x)$, to stand for what may be called participation rates. Thus if $m(x)$ is the age-specific birth rate, the fraction of the births between ages x and $x + dx$ is $be^{-rx}l(x)m(x)\,dx$.

55

This quantity integrated over all ages of positive fertility is $\int_\alpha^\beta be^{-rx}l(x)m(x)\,dx$, where α is the lowest and β the highest age of positive fertility. And this total must give the births per unit of population, so we can equate it to b and obtain the characteristic equation in the unknown r,

$$\int_\alpha^\beta be^{-rx}l(x)m(x)\,dx = b \quad \text{or} \quad \int_\alpha^\beta e^{-rx}l(x)m(x)\,dx = 1. \tag{1}$$

The integrals expressed in terms of their moments enable us to say what is the effect on the intrinsic rate r of variations in the moments of $l(x)m(x)$, known as the net maternity function. Such an argument leads to Chapter IV, on births and stability; the present Chapter III deals only with age distributions related to mortality and natural increase in closed populations.

3.1. Problems

1. The number of individuals between ages x and $x + dx$ in a closed stable population is $be^{-rx}l(x)\,dx$, where b is the overall birth rate, r is the rate of increase, $l(x)$ is the probability at birth of surviving to age x. Give a demonstration in words of this statement.

2. Prove that the constant b in (1) is the reciprocal

$$b = 1 \bigg/ \int_0^\infty e^{-rx}l(x)\,dx,$$

where ω is the highest age to which anyone lives.

3. Show that an addition to the rate of increase has exactly the same effect on age distribution as an addition of the same amount to all age-specific mortality rates.

4. Express the slope of the curve of a stable age distribution in terms of life table functions and the rate of increase.

5. How fast does $\mu(x)$ have to rise with age for the stable age distribution of given rate of increase r to be concave above at age x?

6. The crude death rate in the stable population (also referred to as the intrinsic death rate) is

$$d = \int_0^\infty be^{-rx}l(x)\mu(x)\,dx.$$

Verify that this is equal to $b - r$.

7. A census has been taken of a population assumed to be stable; the count at age x is c_x, and that at age $y > x$ is c_y. Express the rate of increase in terms of c_x, c_y, and the survivorship function $l(x)$.

8. Describe a research project in which one would investigate the accuracy attained by the method used in Problem 7.

9. In a population (that of Mexico 1960) in which the census count at age 25–29 for females was 1,314,000 and at age 50–54 was 538,000, and the number living in the life table for those two ages was 408,000 and 351,000, find the rate of increase.

10. Censuses are often enumerated incorrectly. Suppose that there is an error of fraction δ in the number of individuals c_x enumerated at age x. Find the true rate of increase.

11. In Problem 7 suppose that we use a life table that is appropriate except for the addition to $\mu(x)$ of δ, a random variate of mean zero and variance σ^2. What is the variance of the estimated $\hat{r}$?

12. Show that however far from stable an age distribution is, if the population is closed and the life table is known one can calculate exactly the average rate of increase of births over a past time interval.

13. Generalize the result of Problem 12 to the case where one knows the age distribution in arbitrary intervals, for instance, the number of individuals aged x to $x + u$, ${}_uC_x$, and the number aged y to $y + v$, ${}_vC_y$, where $y > x$.

14. You are told nothing but the mean age of an observed population, and the mean age and variance of the life table that it has followed. With this information estimate its rate of increase.

15. If instead of using the proper life table in Problem 7 we had mistakenly used a life table whose force of mortality $\mu(x)$ was too high by 0.0001 at every age, how far would we have been in error?

16. The case where past birth rates have been constant, but mortality has been steadily improving, is known as quasi-stability (Coale 1963). If the improvement in mortality between successive annual cohorts has been k per year in absolute amount, find how to infer the rate of increase knowing the population count and the life table.

17. If the improvement is in the proportion k, rather than in the absolute amount k that we took in the preceding question, show how to estimate the rate of increase of the population.

18. In a certain population the ratio of male population aged 35–39 to that 60–64 was 2.318, and the corresponding ratio for the life table was 1.249. How fast was the population increasing?

19. Given the following abnormal age distribution for males in Martinique, 1963, use the equation of Problem 7 to calculate the rate of increase with all possible combinations of age.

x	$_5C_x$	$_5L_x$
10	19,118	467,241
20	9,393	461,603
30	8,025	451,857
40	7,194	431,635
50	5,848	389,524
Total—all ages	149,083	

20. As a special case of Problem 7 consider the ratio of girls under 5 to women 15 to 44 at last birthday, all divided by the corresponding ratio for the life table. Show that the result, known as Thompson's index, is an approximation to the net reproduction rate R_0.

21. Show that a better approximation to r than that of Problem 14 is given by

$$\bar{x} = \frac{L_1}{L_0} - \sigma^2 r + \frac{\kappa_3 r^2}{2},$$

where κ_3 is the third moment about the mean of the life table, and obtain an iterative method of solving this. Solve with

$$\frac{L_1}{L_0} = 32.75,$$

$$\sigma^2 = 465.9,$$

$$\kappa_3 = 3585,$$

$$\bar{x} = 31.39$$

(data from France, 1851, females; Keyfitz and Flieger 1968, p. 313).

22. All the conditions of the stable population are met, but there are errors in enumeration in the C_x and C_y such that they are independently subject to normal random variation with mean zero and variance σ^2. To what variation will the rate of increase r as calculated by the expression in Problem 7 be subject?

23. Given the age distribution of a population that can be supposed stable, along with its rate of increase, find its life table.

24. If the life table and the crude death rate of a stable population are known, show how to calculate its rate of increase.

25. How sensitive is the solution in Problem 24 to errors in d?

26. Given the number in the stable population at three ages, find the rate of increase and the life table, on the assumption that the age-specific death rates are fitted by a hyperbola, i.e., that $\mu(x) = \mu_0/(\omega - x)$.

27. Why is it impossible to perform such a fitting if $l(x)$ of the life table is represented by an exponential?

28. A population can be assumed stable, and we know its life table and the fraction of the census population under 25 years of age. Show how to calculate its rate of increase.

29. Show that the proportion of persons in the population aged 60 and over is a declining function of the rate of increase of a population, and express the function in a simple exponential form.

30. With $_{\omega-60}L_{60} = 0.15$ and $m = 25$, tabulate $_{\omega-60}C_{60}$ as a function of r and m_{60} for the values

$$r = 0, 0.01, \ldots, 0.04,$$

$$m_{60} = 67, 68, \ldots, 72.$$

31. Show how to improve the accuracy of the exponential approximation in Problem 29 by incorporating the term in r^2.

32. Show that the derivative of m, the mean age, with respect to r is minus the variance of the distribution $e^{-rx}l(x)\,dx$ taken over the whole range of life, and the derivative of m_{60} is similarly the variance of the ages 60 and over.

33. A retirement colony that people enter at age 70 wants to increase fast enough that its death rate will be no more than 7 percent per year. Given the life table, show how to calculate the required rate of increase.

34. Of two populations, one has a life table $l(x)$, the other $l'(x)$, and the age distribution of the first is $c(x)$, of the second is $c'(x)$. How much of the difference between $c(x)$ and $c'(x)$ is due to mortality differences?

35. What is the interaction of mortality and fertility on the intrinsic rate?

36. Sixty percent of the population of Mexico is under 25 years of age, and we can suppose that in its life table the fraction under 25 is 35 percent, the mean age is 35 and the variance of ages is 550. How fast is Mexico increasing?

37. By how much is the mean age of a population less than that of a standard population, if its rate of increase is greater than the standard by 0.001?

38. Two countries have the same life table, and one is increasing at 2 percent per annum, the other is stationary. The burden of old-age pensions can be defined as the ratio of the population over age 65 to that 20 to 65. What is the difference in the burden between the stationary and the increasing population?

39. Suppose a stable population increasing at 1 percent per year, and with all members subject to a death rate of 1.5 percent uniform at all ages. What fraction will be over 50 years of age?

40. Express the variance of an increasing stable population in terms of the variance and higher cumulants of the life table on which it is based.

41. In a certain hypothetical primitive community men marry at age 40 and women at age 20. How fast does the community have to be increasing for each man to have two wives?

42. Social security benefits of those past 65 paid by the working population year by year, a method called pay-as-you-go, may be contrasted with a funded scheme in which each cohort saves for its old age. Supposing a stable working population, calculate the amount that a person of working age (say 20 to 65) has to pay on the average to support the population over age 65 with $1 per year. Show that this is identical with the actuarial premium on a funded scheme if the rate of interest is equal to the rate of increase of the population.

43. We saw in Chapter I that if there are n_1 births at time t_1, and n_2 births at time t_2, the total number of persons who would have been born in the interval from time t_1 to time t_2 is

$$\text{Persons lived} = Pl = \frac{(n_2 - n_1)(t_2 - t_1)}{\ln n_2 - \ln n_1},$$

on the assumption that the rate of increase of births was constant over the period. Now calculate the number of persons who reached age x (say 60) over the period, and then the number of person-years lived, assuming a fixed life table.

44. Show that at the end of an extended period of geometric increase the fraction of the population still alive for any age is nearly r/b, supposing as before constant rates of mortality and natural increase.

45. In a population with fixed life table and uniformly growing at rate r, how many person-years are lived *by the cohort* born when births are B? How many person-years are lived *in the calendar year* when births are B?

46. Express the relation between the expectation of life and the person-years lived in a given year per one current birth as a simple function of r.

47. How many person-years are lived by the cohorts born between time t_1 when births are b_1 and time t_2 when births are b_2?

48. What about person-years in the calendar years between t_1 and t_2?

49. There are B female births in a particular year in a stable population. How many children will ultimately be born to those females?

50. How many children remain to be born to the N women alive at a particular moment? How many have they already had?

51. Find an expression for the average number of children per year per one member of the stable population, and compare with the corresponding figure for a cohort.

52. Find the sex ratio for persons aged x in the stable population, in which the male and female life tables are different but the rate of increase of the sexes is the same, and the ratio of boy to girl births is s.

53. A commonly used approximation to the stable population in 5-year age intervals is

$$\int_x^{x+5} e^{-ra}l(a)\, da = e^{-(x+2.5)r}\int_x^{x+5} l(a)\, da = e^{-(x+2.5)r}\,_5L_x.$$

Show that this is too low, and find by approximately how much.

54. With a given life table a population increasing rapidly, being young, has a lower crude death rate. But this effect only goes to a certain point; if the rate of increase becomes very great the crude death rate would start to increase again. Explain why in words. With given life table what birth rate makes the death rate a minimum? (Lotka 1938.)

3.2. Solutions

1. Births are increasing at rate r, so x years ago they were e^{-rx} times what they are now. Now they are b per one of total population, so x years ago they must have been be^{-rx}, still per one of present total population. If be^{-rx} is the number of births x years ago, and the fraction of these births that survived is $l(x)$, then the number of survivors must be $be^{-rx}l(x)$. The same argument for a time interval dx rather than one year gives the present number of persons aged x to $x + dx$ as $be^{-rx}l(x)\, dx$.

2. Since $be^{-rx}l(x)\, dx$ is the fraction of the population aged x to $x + dx$ its total over all ages must be unity, i.e.,

$$\int_0^\infty be^{-rx}l(x)\, dx = 1, \quad \text{so } b = \frac{1}{\int_0^\infty e^{-rx}l(x)\, dx}.$$

3. The number of persons at age x is $be^{-rx}l(x)\, dx$, where

$$be^{-rx}l(x)\, dx = be^{-rx}\exp\left(-\int_0^x \mu(a)\, da\right) dx,$$

from which it is plainly indifferent whether one adds a constant δ to the r or to the $\mu(a)$. The same applies to the integral whose reciprocal is b.

4. Differentiating the expression $be^{-rx}l(x)$ with respect to x gives

$$-bre^{-rx}l(x) - be^{-rx}l(x)\mu(x) = -be^{-rx}l(x)(r + \mu(x)).$$

5. The question is equivalent to asking when the second derivative of $be^{-rx}l(x)$ is positive. Twice differentiating

$$e^{-rx}l(x) = \exp\left(-\left(rx + \int_0^x \mu(a)\, da\right)\right)$$

with respect to x gives

$$\frac{d^2(e^{-rx}l(x))}{dx^2} = e^{-rx}l(x)(-\mu'(x) + (r + \mu(x))^2),$$

and the original curve $e^{-rx}l(x)$ is concave above if this is positive, i.e., if $\mu'(x) < (r + \mu(x))^2$.

6. The above definition is equivalent to

$$d = -\int_0^\omega be^{-rx}\, dl(x),$$

and integrating by parts we have

$$-be^{-rx}l(x)\Big|_0^\omega - r\int_0^\omega be^{-rx}l(x)\, dx,$$

or, since the integral on the right is equal to unity, $b - r$. We know that $d = b - r$ for any closed population; the above is merely to check the consistency of our expressions.

7. From stable population theory we have for the fraction of population aged x

$$be^{-rx}l_x = c_x,$$

and for the fraction of population aged y

$$be^{-ry}l_y = c_y,$$

and taking logarithms reduces these to a pair of linear equations of first degree whose solution in $\hat{r}$ is

$$\hat{r} = \frac{1}{y - x} \ln \frac{c_x/l_x}{c_y/l_y}.$$

For this to be a good approximation we need c_x to be the number of persons at age x to nearest birthday, and similarly for c_y. If c_x and c_y are the numbers at last birthday, a better approximation would be obtained by replacing l_x and l_y by

$$L_x \doteq \tfrac{1}{2}(l_x + l_{x+1}) \quad \text{and} \quad L_y \doteq \tfrac{1}{2}(l_y + l_{y+1}).$$

8. At least a lower bound for the variance would be obtained by calculating r from various combinations of x and y. Beyond that one would collect data on a number of populations with known r and $l(x)$, then use the method of Problem 7 to make an estimate $\hat{r}$ for each population, and take the standard error of the departures $\hat{r} - r$.

9. Apply the formula of Problem 7 after adapting it to 5-year ranges to obtain $r = 0.0297$.

10. The true rate of increase, by the same formula as in Problem 7, but with c_x diminished by 100δ percent, is

$$\bar{r} = \frac{1}{y - x} \ln \frac{c_x(1 - \delta)/l_x}{c_y/l_y}$$

$$= \hat{r} + \frac{\ln(1 - \delta)}{y - x} \doteq \hat{r} - \frac{\delta}{y - x}.$$

This suggests that the error in estimating the rate of increase from two ages by relying on stable population theory is inversely proportional to the difference between the ages.

11. We can show that when the life table force of mortality is high by the same δ at all ages the rate of increase as calculated by (7) will be low by δ. It follows that the estimated $\hat{r}$ is subject to variance σ^2.

12. If the population aged x at last birthday is C_x, the number of births x years ago must have been C_x/L_x, and similarly the number of births y years ago must have been C_y/L_y. Hence the average rate of increase of births between x and y years ago must have been the average rate of increase between these numbers:

$$r = \frac{1}{y - x} \ln \frac{C_x/L_x}{C_y/L_y}.$$

The result in Problem 7 applied to a stable population; the present result is general.

13. The equation to be solved for r is now

$$\frac{{}_uC_x}{{}_vC_y} = \frac{\int_x^{x+u} e^{-ra}l(a)\, da}{\int_y^{y+v} e^{-ra}l(a)\, da}.$$

By factoring out $e^{-r(x+u/2)}$ from the integral in the numerator and $e^{-r(y+v/2)}$ from that in the denominator, then solving for the r that is outside the integral signs, we obtain the iterative expression

$$r^* = \frac{1}{y - x + \dfrac{v}{2} - \dfrac{u}{2}} \ln \frac{{}_uC_x \Big/ \displaystyle\int_{-u/2}^{u/2} e^{-rt}l\Big(t + x + \dfrac{u}{2}\Big)\, dt}{{}_vC_y \Big/ \displaystyle\int_{-v/2}^{v/2} e^{-rt}l\Big(t + y + \dfrac{v}{2}\Big)\, dt}.$$

Starting with an arbitrary value of r, say zero, calculating the integrals, thus obtaining r^*, then entering r^* in the integrals, and so proceeding, we will find that convergence occurs with few iterates, especially when x and y are far apart. Note that if $x = y$ no solution is possible.

14. The mean age of the stable population can be written as

$$\frac{\int_0^\infty x e^{-rx} l(x)\, dx}{\int_0^\infty e^{-rx} l(x)\, dx},$$

and each of the integrals can be expanded in an ascending series in r:

$$\bar{x} = \frac{L_1 - rL_2 + r^2 L_3/2! - \cdots}{L_0 - rL_1 + r^2 L_2/2! - \cdots},$$

where

$$L_i = \int_0^\infty x^i l(x)\, dx.$$

Then by ordinary division up to the term in r, we have

$$\bar{x} \doteq \frac{L_1}{L_0} - r\left(\frac{L_2}{L_0} - \left(\frac{L_1}{L_0}\right)^2\right),$$

or if σ^2 is the variance of the stationary age distribution,

$$\bar{x} = \frac{L_1}{L_0} - r\sigma^2.$$

Solving this for r gives the required

$$r = \frac{L_1/L_0 - \bar{x}}{\sigma^2}.$$

15. If the assumed force of mortality $\mu(x)$ was too high by 0.0001, then the true force of mortality was

$$\mu^*(x) = \mu(x) - 0.0001,$$

and the true survivorship was

$$l^*(x) = l(x)e^{0.0001x},$$

and entering this in place of l_x in Problem 7, and similarly for l_y, we end up with

$$\hat{r}^* = \hat{r} + 0.0001$$

in terms of the $\hat{r}$ in (7).

16. If the age-specific rate for those born y years ago is $\mu(a)$, and that for the cohort born x years ago ($x < y$) is $\mu^*(a) = \mu(a) + k(y - x)$, then

$$l_x^* = l_x e^{-k(y-x)x},$$

and the estimate reduces to

$$r_1 = \frac{1}{y - x} \ln \frac{C_x/(l_x e^{-k(y-x)x})}{C_y/l_y}$$

$$= \hat{r} + kx,$$

where $\hat{r}$ is the solution in Problem 7.

17. By the same method, but now with

$$\mu^*(a) = \mu(a)(1 + (y - x)k),$$

the algebra brings us to

$$r_2 = \hat{r} - k \ln l_x.$$

18. The information suffices to apply the expression in Problem 7, so we have

$$r = \frac{1}{62.5 - 37.5} \ln\left(\frac{2.318}{1.249}\right) = 0.0247.$$

19. Our answer to this question is the program in BASIC that follows and provides the numerical results printed out below it.

```
10 'CALCULATION OF RATE OF INCREASE FROM VARIOUS
20 'COMBINATIONS OF AGES
30 FOR I=1 TO 5
40 READ C(I) 'READ IN POPULATION
50 READ L(I) 'READ IN SURVIVORSHIP
60 NEXT I
70 DATA 19118,467241
80 DATA 9393,461603
90 DATA 8025,451857
100 DATA 7194,431635!
110 DATA 5848,389524!
120 'CALCULATE FRACTION AT EACH AGE IN OBSERVED POPULATION
130 FOR I=1 TO 5
140 D(I)=C(I)/149083! 'DIVIDE BY TOTAL POPULATION
150 NEXT I
160 'FIND RATE OF INCREASE FROM ALL COMBINATIONS OF AGES
170 FOR I=1 TO 5
180 FOR J=I+1 TO 5 'INNER LOOP FOR ALL AGE J GREATER THAN I
190 R=D(I)/L(I)/D(J)/L(J)
200 R=LOG (R)
210 R=R/10
220 R=R/(J-I) 'ESTIMATED RATE OF INCREASE
230 'PRINT OUT RESULTS
240 LPRINT 10#I;10#J;
250 LPRINT USING "####.#####";C(I);L(I);C(J);L(J);
260 LPRINT USING "#####.####";R
270 NEXT J,I
280 'EXAMPLE SHOWS VARIABILITY OF ESTIMATE WHEN FORMULA IS
290 'APPLIED TO UNSTABLE POPULATION
```

x	y	$_5C_x$	$_5L_x$	$_5C_y$	$_5L_y$	$\hat{r}$
10	20	19,118	467,241	9,393	461,603	0.0699
10	30	19,118	467,241	8,025	451,857	0.0417
10	40	19,118	467,241	7,194	431,635	0.0299
10	50	19,118	467,241	5,848	389,524	0.0251
20	30	9,393	461,603	8,025	451,857	0.0136
20	40	9,393	461,603	7,194	431,635	0.0100
20	50	9,393	461,603	5,848	389,524	0.0101
30	40	8,025	451,857	7,194	431,635	0.0064
30	50	8,025	451,857	5,848	389,524	0.0084
40	50	7,194	431,635	5,848	389,524	0.0104

The considerable variation in the resultant value of $\hat{r}$, shown in the last column, is typical of populations that are both imperfectly enumerated and in process of destabilization as their birth and death rates fall.

20. The expression for Thompson's index may be written

$$\frac{{}_5C_0}{{}_{30}C_{15}} \bigg/ \frac{{}_5L_0}{{}_{30}L_{15}}.$$

The integral

$$_5C_0 = \int_0^5 e^{-rx}l(x)\,dx$$

is approximately

$$e^{-2.5r}\int_0^5 l(x)\,dx = e^{-2.5r}\,{}_5L_0,$$

and similarly

$$_{30}C_{15} = \int_{15}^{45} e^{-rx}l(x)\,dx \doteq e^{-30r}\int_{15}^{45} l(x)\,dx,$$

so Thompson's index reduces to $e^{27.5r}$, and insofar as $27\frac{1}{2}$ years is close to the length of a generation this is the ratio of increase over a generation, or R_0.

21. The expression in the question comes from taking the synthetic division one term beyond that of Problem 14. To solve we transpose and write the form

$$r^* = \frac{\dfrac{L_1}{L_0} - \bar{x}}{\sigma^2 - \dfrac{\kappa_3\, r}{2}},$$

which is identical with the equation to be solved if $r = r^*$. Putting an asterisk beside the r on the left reminds us that it is an improved value until convergence is attained.

```
10 'ITERATIVE SOLUTION OF QUADRATIC EQUATION
20 L=32.75:S2=465.9:'ENTER VALUES OF COEFFICIENTS
30 K3=3585:X=31.39
40 R=.01 'INITIAL VALUE FOR ITERATION
50 'ITERATE UNTIL SUCCESSIVE VALUES AGREE TO 7 PLACES
60 WHILE R-R1 'DIFFERENCE OF SUCCESSIVE VALUES
70 R=R1 'REPLACE FIRST ITERATE BY SECOND
80 R1=L-X
90 R1=R1/(S2-K3*R/2) 'CALCULATION OF NEXT ITERATE
100 LPRINT USING "###.#####";R1
110 WEND 'LOOP BACK TO 60 UNTIL R=R1
```

$$0.00292$$
$$0.00295$$
$$0.00295$$
$$0.00295$$
$$0.00295$$

22. Using the formula of Solution 7 we need to know how r will be affected by the addition of random δ to C_x and ε to C_y. If we subtract the random errors we would obtain the true $\hat{r}_v$, say:

$$\hat{r}_v = \frac{1}{y - x} \ln \frac{(C_x - \delta)/l_x}{(C_y - \varepsilon)/l_y},$$

or

$$\hat{r}_v = \frac{1}{y - x} \ln\left[\left(\frac{C_x}{l_x}\right)\left(1 - \frac{\delta}{C_x}\right)\right] - \ln\left[\left(\frac{C_y}{l_y}\right)\left(1 - \frac{\varepsilon}{C_y}\right)\right]$$

$$= \hat{r} + \ln\left(\frac{1 - \dfrac{\delta}{C_x}}{1 - \dfrac{\varepsilon}{C_y}}\right)\Bigg/(y - x),$$

where $\hat{r}$ is the solution in No. 7. Then

$$\hat{r}_v \doteq \hat{r} - \left(\frac{\delta}{C_x} - \frac{\varepsilon}{C_y}\right)\Bigg/(y - x)$$

if the perturbations are small. And since they are uncorrelated the variance of $\hat{r}_v$ will be $(\sigma^2/C_x^2 + \sigma^2/C_y^2)/(y - x)^2$.

23. The fraction of the population aged x is $be^{-rx}l(x)$; if the number of individuals enumerated at x is the known $c(x)$ we have the equation

$$c(x) = be^{-rx}l(x), \qquad x = 0, 1, 2, \ldots$$

to be solved for each value of x.

24. The birth rate b is equal to

$$1\Bigg/\int_0^\omega e^{-rx}l(x)\,dx.$$

But the birth rate is equal to the death rate d plus the rate of natural increase r. Hence we have the equation

$$d + r = \frac{1}{\int_0^\omega e^{-rx}l(x)\,dx},$$

of which the only unknown is r, solvable by iteration.

25.

$$\frac{dr}{dd} = \left\{\frac{\int_0^\omega xe^{-rx}l(x)\,dx}{[\int_0^\omega e^{-rx}l(x)\,dx]^2} - 1\right\}^{-1}.$$

26. If $\mu(x) = \mu_0/(\omega - x)$ then by integration and taking the exponential,

$$l(x) = \left(1 - \frac{x}{\omega}\right)^{\mu_0}.$$

The equation to be fitted is

$$C_x = be^{-rx}l(x), \qquad x = x_1, x_2, x_3,$$

where the C_x are known. With the hyperbolic life table this becomes

$$C_x = be^{-rx}\left(1 - \frac{x}{\omega}\right)^{\mu_0},$$

so we have three equations for the three unknowns b, r, and μ_0, taking ω as 100 or other arbitrary number.

27. If the life table were $e^{-\lambda x}$ we would face a problem of identification; since the rate of increase and the constant enter only as the sum $r + \lambda$, no amount of data will enable us to ascertain their separate values.

28. The fraction of the population under 25 years of age is

$$\alpha = \frac{\int_0^{25} e^{-rx}l(x)\,dx}{\int_0^{\omega} e^{-rx}l(x)\,dx}.$$

This is the required equation for the unknown intrinsic rate r. For an iterative formula that will converge to r, multiply both sides by e^{10r} and rearrange to obtain

$$r^* = \frac{1}{10}\ln\left[\frac{\alpha \int_0^{\omega} e^{-r(x-20)}l(x)\,dx}{\int_0^{25} e^{-r(x-10)}l(x)\,dx}\right].$$

Converting to the usual life table functions in 5-year intervals gives a form suitable for calculation:

$$r^* = \frac{1}{10}\ln\left[\frac{\alpha \displaystyle\sum_{i=0}^{i=\omega-5} e^{-r(i-17.5)}\,{}_5L_i}{\displaystyle\sum_{i=0}^{20} e^{-r(i-7.5)}\,{}_5L_i}\right],$$

where the summation is over every fifth age.

29. The proportion of the stable population aged x to $x + dx$ is

$$be^{-rx}l(x)\,dx,$$

where $l(x)$ is the probability of surviving from birth to age x, r is the rate of increase, and b is the birth rate. Then the proportion aged 60 and over is

$$_{\omega-60}C_{60} = \frac{\int_{60}^{\omega} be^{-rx}l(x)\,dx}{\int_0^{\omega} be^{-rx}l(x)\,dx},$$

where ω is the highest age to which anyone lives. By taking logarithms, then differentiating:

$$\frac{d \ln {}_{\omega-60}C_{60}}{dr} = \frac{d {}_{\omega-60}C_{60}}{{}_{\omega-60}C_{60}\, dr} = -\frac{\int_{60}^{\omega} x e^{-rx}l(x)\, dx}{\int_{60}^{\omega} e^{-rx}l(x)\, dx} + \frac{\int_{0}^{\omega} x e^{-rx}l(x)\, dx}{\int_{0}^{\omega} e^{-rx}l(x)\, dx}$$

$$= m - m_{60}, \tag{1}$$

where m is the mean age of the whole population, and m_{60} is the mean age of those 60 and over. Since m is necessarily smaller than m_{60} the derivative is negative. This proves that the logarithm of the proportion over age 60 is a declining function of r, and hence that the proportion itself declines with r.

To approximate the function, note that

$$\frac{d \ln {}_{\omega-60}C_{60}}{dr} = m - m_{60}$$

can be considered a differential equation, whose solution is

$$\ln {}_{\omega-60}C_{60} = (m - m_{60})r + \kappa,$$

κ being an arbitrary constant, or if $L = e^{\kappa}$

$$_{\omega-60}C_{60} = Le^{(m-m_{60})r}.$$

We would like to choose the constant L so that when r is zero the proportion is that of the life table. This gives in the end

$$_{\omega-60}C_{60} \doteq \frac{{}_{\omega-60}L_{60}}{\mathring{e}_0}\, e^{(m-m_{60})r}, \tag{2}$$

$\mathring{e}_0$ being the expectation of life at age zero, and the L function the integral of the $l(x)$ from age 60 to the end of life.

As a numerical example applicable to a rapidly growing population, r might be 0.03, m might be 25 years of age, m_{60} might be 70 years, and the proportion over 60 in the life table might be 0.15. Then the proportion over 60 in the stable population is

$$_{\omega-60}C_{60} \doteq 0.15e^{-45r} = 0.15e^{-1.35} = 0.0389.$$

30.

```
1 'PROPORTION OVER 60 IN THE STABLE POPULATION
2 FOR M60=67 TO 72 'OUTER LOOP FOR VALUES OF MEAN AGE
3 LPRINT M60;
4 FOR R=0 TO .04 STEP .01 'INNER LOOP FOR VALUES OF RATE  R
5 PROP=.15*EXP((25-M60)*R) 'CALCULATE PROPORTION OVER 60
6 LPRINT USING "#####.####";PROP,
7 NEXT R 'END INNER LOOP
8 LPRINT
9 NEXT M60 'END OUTER LOOP
```

Mean age of pop 60+	Values of r				
	0	0.01	0.02	0.03	0.04
67	0.1500	0.0986	0.0648	0.0425	0.0280
68	0.1500	0.0976	0.0635	0.0413	0.0269
69	0.1500	0.0966	0.0622	0.0401	0.0258
70	0.1500	0.0956	0.0610	0.0389	0.0248
71	0.1500	0.0947	0.0598	0.0377	0.0238
72	0.1500	0.0938	0.0586	0.0366	0.0229

The conclusion is that the proportion varies moderately with the mean age assumed and a great deal with the rate of increase.

31. We have the expansion of $\ln {}_{\omega-60}C_{60}$ as $\kappa + (m - m_{60})r + (\sigma_{60}^2 - \sigma^2)(r^2/2)$, where κ as before is the logarithm of ${}_{\omega-60}L_{60}/\mathring{e}_0$, σ_{60}^2 is the variance in age of the population over 60 and σ^2 the variance in age of the entire population. Taking exponentials gives

$$_{\omega-60}C_{60} \doteq \frac{{}_{\omega-60}L_{60}}{\mathring{e}_0} \exp((m - m_{60})r + (\sigma_{60}^2 - \sigma^2)(r^2/2)).$$

Note that the differentiations, integrations, and functions here are with respect to populations having different values of r, conceived as existing contemporaneously. The argument belongs to what is called comparative statics, in which various stable conditions are compared.

32. By definition

$$m = \frac{\int_0^\omega xe^{-rx}l(x)\,dx}{\int_0^\omega e^{-rx}l(x)\,dx},$$

$$\frac{dm}{dr} = \frac{-\int_0^\omega x^2 e^{-rx}l(x)\,dx}{\int_0^\omega e^{-rx}l(x)\,dx} + \left(\frac{\int_0^\omega xe^{-rx}l(x)\,dx}{\int_0^\omega e^{-rx}l(x)\,dx}\right)^2,$$

and this is minus the variance of the age distribution through the whole of life. The argument carries through unchanged if the lower limit of the integral is 60 throughout, or any other age between 0 and ω.

33. The death rate in the colony is

$$\frac{\int_{70}^\omega e^{-rx}l(x)\mu(x)\,dx}{\int_{70}^\omega e^{-rx}l(x)\,dx} \tag{1}$$

and we need to equate this to 0.07 and then solve for r. For an iterative solution we multiply both sides of the equation by e^{5r} and both integrals by e^{75r} to obtain

$$\frac{\int_{70}^\omega e^{-r(x-80)}l(x)\mu(x)\,dx}{\int_{70}^\omega e^{-r(x-75)}l(x)\,dx} = 0.07e^{5r},$$

and if r^* is the improved value in each cycle of iteration this becomes on rearrangement

$$r^* = \frac{1}{5}\left[\ln\left[\frac{\int_{70}^{\omega} e^{-r(x-80)}l(x)\mu(x)\,dx}{\int_{70}^{\omega} e^{-r(x-75)}l(x)\,dx}\right] - \ln 0.07\right].$$

A numerical example for Canada 1965 females follows. Data are the $_5d_x$ and $_5L_x$ columns of the life table.

```
10 'ITERATIVE PROCEDURE FOR FINDING
20 'RATE OF INCREASE FOR RETIREMENT HOME
30 'DATA FOR CANADA FEMALES 1965
40 'FROM KEYFITZ & FLIEGER, 1968, P. 90
50 'D(I) AND L(I) ARE LIFE TABLE DEATHS AND
60 'NUMBER LIVING RESPECTIVELY
70 FOR I=1 TO 4
80 READ D(I),L(I)
90 NEXT I
100 DATA 11490,343248
110 DATA 15150,276570
120 DATA 18504,191061
130 DATA 28943,146959
140 R1=.01 'ARBITRARY INITIAL VALUE OF RATE OF INCREASE
150 WHILE R1-R
160 'BUILD UP NUMERATOR AND DENOMINATOR OF LOGARITHM
170 R=R1
180 N=0
190 D=0
200 FOR I=1 TO 4
210 N=N+D(I)*EXP((12.5-5*I)*R) 'NUMERATOR
220 D=D+L(I)*EXP((7.5-5*I)*R) 'DENOMINATOR
230 NEXT I
240 'NOW THE ITERATIVE PROCESS PROPER
250 R1=(LOG(N/D)-LOG(.07))/5 'IMPROVED VALUE OF RATE OF INCREASE
260 LPRINT R1,
270 WEND 'LOOP BACK TO "WHILE"
280 LPRINT: LPRINT
290 LPRINT "THE HOME WOULD HAVE TO INCREASE AT";
300 LPRINT 100*R1;"PERCENT PER YEAR"
310 END
```

```
.0227015      .0262595      .0272797      .0275745      .0276599
.0276947      .0276947

THE HOME WOULD HAVE TO INCREASE AT 2.76947 PERCENT PER YEAR
```

34. Writing out the expressions for $c(x)$ and for $c'(x)$ and subtracting gives for the overall difference

$$c'(x) - c(x) = \frac{e^{-r'x}l'(x)}{\int_0^{\omega} e^{-r'x}l'(x)\,dx} - \frac{e^{-rx}l(x)}{\int_0^{\omega} e^{-rx}l(x)\,dx},$$

where r is the solution of $\int_{\alpha}^{\beta} e^{-rx}l(x)m(x)\,dx = 1$ and r' is the solution of $\int_{\alpha}^{\beta} e^{-r'x}l'(x)m'(x)\,dx = 1$. The part of this attributable to mortality is obtained by the corresponding difference with fertility made the same for the two populations, that is, we would take the above difference except for substituting in the first term an r obtained as the solution of

$$\int_{\alpha}^{\beta} e^{-rx}l'(x)m(x)\,dx = 1,$$

say $r(l', m)$, and in the second $r(l, m)$. Similarly for the effect of fertility differences, which would require us to solve the characteristic equation with $m'(x)$ and $l(x)$.

35. In the notation of Solution 34,

$$r(l', m') - r(l, m') - r(l', m) + r(l, m).$$

36. Filling in what we know in the formula of Solution 31, suitably modified to represent under 25 rather than 60 and over, we have

$$0.60 = 0.35e^{(35-m_{-25})r + (\sigma^2_{-25} - 550)(r^2/2)},$$

where m_{-25} is the mean age of those under 25 and σ^2_{-25} their variance. But the mean age of the under 25 cannot be very far from 10, and the variance of the ages under 25 cannot be far from the square of one-third of the interval, i.e., about 60. Taking logarithms this gives an equation in which only r is missing:

$$\ln\left(\frac{0.60}{0.35}\right) = (35 - 10)r + (60 - 550)(r^2/2)$$

or

$$0.539 = 25r - 245r^2.$$

Such an equation may be solved by the quadratic formula, or if a programmable calculator is available by rearrangement to obtain the iterative process:

$$r^* = \frac{0.539}{25 - 245r}.$$

Successive iterates are 0, 0.0216, 0.0273, 0.0294, 0.0303, 0.0307, 0.0308, 0.0309.

Following is a BASIC program for accomplishing this.

```
10 'ITERATIVE PROCESS FOR THE INTRINSIC RATE
20 'A METHOD FOR SOLVING A QUADRATIC EQUATION
30 R=.01 'INITIAL VALUE OF RATE
40 WHILE R1-R '
50 R=R1
60 R1=.539/(25-245*R) 'IMPROVED VALUE
70 LPRINT R;
80 WEND 'BACK TO "WHILE"
```

```
0   .02156    .0273357    .0294491    .0303065    .0306687    .0308244    .0308917
.0309434    .0309434    .0309435    .0309435    .0309435    .0309435
```

37. Using the expression developed in Solution 14 the answer must be that a population higher in rate of increase by 0.001 must have a mean age lower by $0.001\sigma^2$, where σ^2 is the variance of the life table. The typical life table has variance about 500; for example, the United States in 1964 showed 500.0 for males and 560.5 for females (Keyfitz and Flieger 1968, p. 175). Hence one can say that the faster growing population will have a mean age lower by about 0.001(500), or half a year.

38. Applying the expression for the burden

$$b = \frac{\int_{65}^{\omega} e^{-rx} l(x)\, dx}{\int_{20}^{65} e^{-rx} l(x)\, dx}$$

which is derived from the stable population, taking logarithms of both sides, and differentiating, we have

$$\frac{d \ln b}{dr} = \kappa_1 - k_1,$$

where κ_1 is the mean of the population 20 to 65, k_1 the mean of those over 65. In finite terms, and supposing $\kappa_1 = 35$, $k_1 = 75$, we have

$$\frac{\Delta b}{b} = (\kappa_1 - k_1)\Delta r = -40\Delta r,$$

so that for each percentage point of difference in r there will be a difference of 40 percent (*not* percentage points) in b in the opposite direction.

39. The fraction must be

$$f = \frac{\int_{50}^{\infty} e^{-rx} l(x)\, dx}{\int_{0}^{\infty} e^{-rx} l(x)\, dx} = \frac{\int_{50}^{\infty} e^{-(0.01 + 0.015)x}\, dx}{\int_{0}^{\infty} e^{-(0.01 + 0.015)x}\, dx}$$

$$= e^{-(0.025)(50)} = e^{-1.25} = 0.287.$$

40. We need to estimate

$$\sigma_s^2 = \frac{\int_0^{\omega} (x - \bar{x})^2 e^{-rx} l(x)\, dx}{\int_0^{\omega} e^{-rx} l(x)\, dx},$$

by expanding its exponentials and then dividing.

41. If the population is stable, and men and women increase at the same rate, and if equal numbers of boy and girl babies are born, then the number of men aged 40 will be $Be^{-40r} l_{40}^*$ and of women aged 20 $Be^{-20r} l_{20}$, where B is the number of current births, and the male life table is distinguished with a star. The conditions of the problem may be expressed as

$$Be^{-20r} l_{20} = 2Be^{-40r} l_{40}^*,$$

so we must have

$$e^{20r} = \frac{2 l_{40}^*}{l_{20}},$$

and therefore

$$r = \frac{1}{20} \ln \frac{2 l_{40}^*}{l_{20}}.$$

For Togo 1961, l_{40}^* for males is 0.43, l_{20} for females is 0.62 (Keyfitz and Flieger 1968, p. 74), so we have the required r as 0.016. The very moderate increase of 1.6 percent per year suffices for each man to have two wives.

42. The pay-as-you-go (payg) premium is the ratio of the population of drawing age to that of working age, i.e.,

$$P_{\text{payg}} = \frac{\int_{65}^{\omega} e^{-rx} l(x)\, dx}{\int_{20}^{65} e^{-rx} l(x)\, dx}.$$

For the funded scheme, if we discount to birth the premiums and the benefits at rate δ,

$$P_{\text{fund}} = \frac{\int_{65}^{\omega} e^{-\delta x} l(x)\, dx}{\int_{20}^{65} e^{-\delta x} l(x)\, dx}.$$

If $\delta = r$ the two are the same.

43. The initial population aged x would have been $n_1 e^{-rx} l(x)$, and the final population $n_2 e^{-rx} l(x)$, and so with geometric increase we would have for the total attaining age x

$$Pl_x = \frac{(n_2 e^{-rx} l(x) - n_1 e^{-rx} l(x))(t_2 - t_1)}{\ln n_2 - \ln n_1}$$

$$= (e^{-rx} l(x))(Pl),$$

Pl being the person-years lived by all ages together. In short, we need only multiply the total lived by the number aged x as a fraction of births. Total person-years lived would be

$$Pyl = Pl \int_0^{\omega} e^{-rx} l(x)\, dx = Pl/b.$$

where b is the crude or overall birth rate.

44. If at the start the number of births is 1, then the number of persons who lived is nearly

$$\frac{(n_2 - n_1)(t_2 - t_1)}{\ln n_2 - \ln n_1} \doteq \frac{(n_2)(t_2)}{\ln(e^{rt_2})} = \frac{n_2}{r},$$

i.e., number of births divided by rate of increase. Then dividing by the population at time t_2, this becomes b/r, where b is the birth rate. The ratio of the current population to the ever-lived is the reciprocal of this, or r/b.

45. The cohort lives $B\mathring{e}_0$ person-years; the person-years lived in the calendar year are

$$B \int_0^{\omega} e^{-rx} l(x)\, dx$$

or B/b, where b is the crude or intrinsic birth rate.

46. For the expectation of life we have

$$\mathring{e}_0 = \int_0^\omega l(x)\, dx$$

and for the person-years in a given calendar year we have

$$\int_0^\omega e^{-rx} l(x)\, dx.$$

To expand the integral in the latter expression by developing e^{-rx} in powers of x would not do because the moments about any point become very large. But the cumulants increase slowly, so we take the logarithm of $\int_0^\omega e^{-rx} l(x)\, dx$ to obtain

$$\ln \frac{\int_0^\omega e^{-rx} l(x)\, dx}{\mathring{e}_0} \doteq -\frac{L_1}{L_0} r + \sigma^2 \frac{r^2}{2},$$

$$\frac{\int_0^\omega e^{-rx} l(x)\, dx}{\mathring{e}_0} = \exp\!\left(-\frac{L_1}{L_0} r + \sigma^2 \frac{r^2}{2} \right)$$

$$\doteq 1 - \frac{L_1}{L_0} r + \sigma^2 \frac{r^2}{2},$$

or

$$\int_0^\omega e^{-rx} l(x)\, dx \doteq \mathring{e}_0 \!\left(1 - \frac{L_1}{L_0} r + \sigma^2 \frac{r^2}{2} \right),$$

so the required difference is nearly

$$\mathring{e}_0 \!\left(\frac{L_1}{L_0} r - \frac{\sigma^2 r^2}{2} \right).$$

The relation between the two quantities is easily found in terms of life table moments and the rate of increase r.

47. The total births during the time are given in Chapter I as

$$\frac{(t_2 - t_1)(b_2 - b_1)}{\ln b_2 - \ln b_1};$$

the total person-years are this times $\mathring{e}_0$.

48. Same as before, but multiplying by $\int_0^\omega e^{-rx} l(x)\, dx$ instead of $\mathring{e}_0$.

49. The probability of each of the girl children reaching x years is $l(x)$; the probability of her having a child between age x and $x + dx$ is $m(x)$, say; the total expected children that will be born to her is the sum

$$\int_0^\omega l(x) m(x)\, dx,$$

a quantity known as the net reproduction rate and designated R_0. The total number of children expected to be born is BR_0.

50. Among the N women are $Nbe^{-ra}l(a)\,da$ aged a to $a + da$. Women a years of age can expect to have

$$\int_a^\beta l(x)m(x)\,dx$$

children; the total expected must therefore be

$$Nb \int_0^\beta e^{-ra}l(a) \int_a^\beta l(x)m(x)\,dx\,da,$$

where α is the youngest and β the oldest age of childbearing. The number of children they have already had differs in that the inner integral has limits α and a, *and* omits the $l(x)$, since those women are alive at the time:

$$Nb \int_0^\beta e^{-ra}l(a) \int_\alpha^a m(x)\,dx\,da.$$

Note that the number they have had plus the number they will have add to more than the total NR_0 obtained by assigning R_0 to each.

51. We have $b = 1/\int_0^\omega e^{-rx}l(x)\,dx$ for the stable population and $b' = 1/\int_0^\omega l(x)\,dx$ for the cohort.

52. Per one girl born the number of females aged x is $e^{-rx}l(x)$; the number of boy births is s, and hence the number of males aged x is $se^{-rx}l*(x)$, where $l*(x)$ is the male survivorship function and $l(x)$ the female. Hence the ratio of the male to the female population at age x is

$$\frac{se^{-rx}l*(x)}{e^{-rx}l(x)} = \frac{sl*(x)}{l(x)},$$

a quantity dependent only on the life table and the sex ratio at birth, but not on the common rate of increase.

53. Put $a = u + x + 2\frac{1}{2}$. Then

$$\int_x^{x+5} e^{-ra}l(a)\,da = e^{-r(x+2.5)}\left({}_5L_x + \frac{r^2}{2} \int_{-2.5}^{2.5} u^2 l(u + x + 2\tfrac{1}{2})\,du\right),$$

supposing $l(a)$ is close to a straight line (or other odd function) within the age interval. If this is so the coefficient of r^2 is approximately equal to the variance σ_x^2 of $l(a)$ within the interval $x, x + 5$, so we have

$$\int_x^{x+5} e^{-ra}l(a)\,da = e^{-r(x+2.5)}\left({}_5L_x + \frac{r^2\sigma_x^2}{2}\right).$$

54. We want to find the minimum of

$$d = \frac{\int_0^\omega e^{-rx}l(x)\mu(x)\,dx}{\int_0^\omega e^{-rx}l(x)\,dx}.$$

Using the fact that $l(x)\mu(x)\,dx = -dl(x)$ and integrating in the numerator by parts, we have

$$d = \frac{1}{\int_0^\infty e^{-rx}l(x)\,dx} - r.$$

To find the minimum differentiate with respect to r and equate to zero to obtain

$$b\bar{A}_r - 1 = 0,$$

where $\bar{A}_r$ is the mean age in the stable population of rate of increase r. The solution is $b = 1/\bar{A}_r$; iteration is required to find the precise answer.

CHAPTER IV
Births and Deaths Under Stability

The previous chapter dealt with the properties of a population having a given life table and an arbitrary rate of increase r. In this chapter the rate of increase is no longer arbitrary, but determined by the life table and age-specific birth rates. We start with several wholly unoriginal questions designed only to elicit the basic theory.

4.1. Problems

1. Set down an equation showing the number of births in one generation in terms of the number of births in the preceding generation, age-specific rates of birth and death being given and fixed. Confine the model to one sex, say females.

2. Solve the equation of Problem 1 for the implied rate of increase on the supposition that the trajectory is an exponential starting with B_0 at time zero.

3. Sketch any one of the methods available for solving the equation of Problem 2 for the unknown r, given the net maternity function $l(x)m(x)$.

4. Given the net maternity function for mothers and girl babies, $_5L_x {}_5F_x$, of Ireland in 1926, find the intrinsic rate of natural increase r. (Keyfitz and Flieger 1968, p. 380.)

x	$_5L_x$	$_5F_x$	$_5L_x \, _5F_x$
15	4.321	0.0023	0.0098
20	4.232	0.0347	0.1468
25	4.127	0.0805	0.3322
30	4.011	0.0897	0.3597
35	3.882	0.0660	0.2562
40	3.740	0.0324	0.1211

5. Given similar data for Mexico, 1962, find Mexico's rate of increase, and show how much of the difference between Ireland's rate of increase and Mexico's is due to differences in births, how much to differences in deaths.

x	$_5L_x$	$_5F_x$
15	4.344	0.2138
20	4.296	0.6171
25	4.230	0.6538
30	4.256	0.5113
35	4.061	0.4123
40	3.955	0.1865

6. It is shown in textbooks of numerical analysis, for example, Scarborough (1958, p. 209), that the convergence of a functional iteration such as that of the preceding problem depends only on the derivative dr^*/dr being less than unity. If it is much less, convergence will be rapid. Find the derivative in the case of the formula for the intrinsic rate of Problem 3, and estimate roughly the number of decimal points obtained on each iteration.

7. Find the first two derivatives of the function, which may be designated $\psi(r)$, on the right-hand side of Problem 2, and state what their signs are.

8. Trace the function $\psi(r)$ in Problem 2 and its first two derivatives as calculated in Problem 7.

9. Prove that the equation in Problem 2 can have only one real root.

10. Apply the method of Problem 29 of Chapter III to births and show that the proportion of births to women over 30 varies rather little with the rate of increase.

11. Derive

$$\frac{\sigma^2 r^2}{2} - mr + \ln R_0 = 0, \tag{1}$$

where R_0 is the integral of the net maternity function and m and σ^2 its mean and variance, as an approximation to the characteristic equation

$$\int_\alpha^\beta e^{-rx}l(x)m(x)\, dx = 1.$$

12. If two stable populations have identical age-specific birth rates, and mortality rates in one are higher by δ at every age, show that their age-distributions are identical.

13. Show that a change of 0.001 (or any other amount) in death rates at all ages leaves age distributions unchanged for the short run, irrespective of whether the population is stable or not.

14. What difference does it make to the intrinsic rate if the age-specific fertility rate is lower by a small fraction f at every age?

15. If the death rate at all ages is lower by δ in one population compared with another, by how much must the birth rate be lower to offset this so that the rate of increase of both populations is the same?

16. Show that with $r = 0$ in Problem 2 each birth is replaced in the next generation by just one birth.

17. If the force of mortality is increased uniformly at all ages, and all else remains constant, the intrinsic rate will decrease. Show that an increase in the force of mortality equal to the intrinsic rate would ultimately bring the population down to stationarity.

18. Call the probability of X falling between x and $x + dx$ $f(x)\, dx$, the mean value of X

$$\bar{x} = \mu_1' = \int_{-\infty}^{\infty} x f(x)\, dx,$$

and the series of moments about zero,

$$\mu_i' = \int_{-\infty}^{\infty} x^i f(x)\, dx.$$

Then cumulants κ_i are the coefficients of the expansion of the logarithm of the moment-generating function, i.e.,

$$\ln\left(\int_{-\infty}^{\infty} e^{rx} f(x)\, dx\right) = \kappa_1 r + \frac{\kappa_2 r^2}{2!} + \frac{\kappa_3 r^3}{3!} + \cdots.$$

Express κ_1 and κ_2 in terms of μ_1' and μ_2'.

19. The length of generation is defined as that time in which a population increasing at annual rate r compounded momently will grow in the ratio of the net reproduction rate, i.e., it is defined by the equation $R_0 = e^{rT}$. Show how T can be expressed as an infinite series in r.

20. Prove that the length of generation is the average of the mean age of childbearing in the stationary and in the stable populations, approximately.

21. How tight are the bounds that can be set on r when R_0 is given?

22. If one quarter of the women of a country leave at the age of 35, write the equation that will show what that does to the rate of increase.

23. Show that the intrinsic rate is positive when R_0 is greater than unity; zero when $R_0 = 1$; negative when R_0 is less than unity.

24. A woman has a daughter at the age of 23 and another at the age of 30. What would be the intrinsic rate of a population in which all women who survived had exactly this childbearing pattern, given $l_{23} = 0.95$ and $l_{30} = 0.93$?

25. What happens to the characteristic equation in Problem 2 in the special case where all cumulants beyond the second are zero?

26. By how much does a decrease of one year in the mean age of child-bearing raise the intrinsic rate?

27. Check the derivatives of the preceding question numerically.

28. What is the average age at death in a stable population with force of mortality μ the same at all ages and rate of increase r? Contrast this result with the average age at which an individual can expect to die.

29. If of two stable populations one has death rates at all ages higher by δ than the other in absolute amount, and is increasing less rapidly by δ, show that the two have identical birth rates.

30. Prove the converse of the above: that if two stable populations have identical crude birth rates and one has mortality at all ages higher by δ than the other, then the one with the higher death rates increases less rapidly by δ.

31. A population is heterogeneous, containing two equal subpopulations, both initially stable, equal in number and increasing at the same rate r, but with different life tables. A medical procedure is applied that lowers mortality by amount ε at all ages for one of the subpopulations. Under what circumstances will the use of this procedure *raise* the crude death rate of the population?

32. If $\omega(t)$ is an increasing function, and $a < 1$, show that

$$s(t) = \frac{\omega(t)\,a + 1}{\omega(t) + 1}$$

is a decreasing function.

33. Show that in a heterogeneous population a medical improvement can diminish the chances of survival for a randomly chosen individual.

4.2. Solutions

1. We seek the unknown births as a function of time $B(t)$. From the age-specific death rates or force of mortality $\mu(x)$ a life table can be calculated, with $l(x)$ as the probability of survival from birth to age x. The number of women aged x at time t must be equal to the births x years before, $B(t - x)$, less deaths, i.e., $B(t - x)l(x)$. The birth rate being $m(x)\,dx$ for the age interval x to $x + dx$, the women produce $B(t - x)l(x)m(x)\,dx$ children. The total of these through all ages must be equal to the current number of births $B(t)$, so we have the equation

$$B(t) = \int_\alpha^\beta B(t - x)l(x)m(x)\,dx,$$

where α is the youngest age of childbearing and β the oldest. This equation does not permit a solution for the birth trajectory, $B(t)$, in absolute numbers, since it does not provide any starting point for the process, but gives each generation in terms of the preceding one.

2. If the unknown trajectory is $B_0\,e^{rt}$ we can obtain an equation for the constant r in place of the far more difficult equation for the function $B(t)$. Entering

$$B(t) = B_0\,e^{rt}$$

in the equation relating two successive generations, and then cancelling out $B_0\,e^{rt}$ from both sides, we have

$$1 = \int_\alpha^\beta e^{-rx}l(x)m(x)\,dx \tag{2}$$

as the equation for r.

3. Multiply both sides by $e^{27.5r}$, take logarithms, and divide by $27\frac{1}{2}$. This gives

$$r^* = \frac{1}{27\frac{1}{2}}\ln\int_\alpha^\beta e^{-r(x-27.5)}l(x)m(x)\,dx,$$

where the r on the left has been starred to show that at intermediate stages it is an improved value. Once convergence has been attained $r = r^*$, and when that is true r must satisfy the original equation of Solution 2.

If one enters an arbitrary starting value for r on the right-hand side, say zero, that gives an improved value on the left, r^*, which can in turn be entered on the right, and so continuing. Usually three or four iterations provide all the accuracy needed. In practice one would enter a sum of 5-year age groups on the right in place of the integral; the $27\frac{1}{2}$ has been arbitrarily chosen as the center of one of these groups near the mean age of childbearing.

4. The solution is as follows, expressed in the MBASIC language:

```
10 'FUNCTIONAL ITERATION FOR THE INTRINSIC RATE
20  'FIRST READ IN THE LIFE TABLE, THEN THE
30 'FEMALE AGE-SPECIFIC FERTILITY RATES
40 FOR I=1 TO 6
50 READ LL(I)
60 LL(I)=LL(I)/1000
```

```
70 NEXT I
80 FOR I=1 TO 6
90 READ F(I)
100 F(I)=F(I)/10000
110 N(I)=LL(I)*F(I)
120 NEXT I
130 DATA 4321,4232,4127,4011,3882,3740
140 DATA 23,347,805,897,660,324
150 R=.01
160 'THE LOOP IS DESIGNED TO CONTINUE UNTIL RSTAR-R=0
170 WHILE RSTAR-R
180 R=RSTAR
190 A=0
200 FOR I=1 TO 6
210 A=A+EXP((15-5*I)*R)*N(I)
220 NEXT I
230 RSTAR=LOG(A)/27.5
240 PRINT USING "#####.#####";RSTAR
250 WEND
260 END
```

and this produced the sequence:

0.00741
0.00627
0.00645
0.00642
0.00642
0.00642

5. By converting the iteration into a subroutine and permuting the mortality and fertility schedules, we obtain the printout below:

```
1 'FUNCTIONAL ITERATION FOR THE INTRINSIC RATE
2  'FIRST READ IN THE LIFE TABLE, THEN THE
3 'FEMALE AGE-SPECIFIC FERTILITY RATES
4 FOR I=1 TO 6
5 READ LL(I)
6 LL(I)=LL(I)/1000
7 NEXT I
8 FOR I=1 TO 6
9 READ F(I)
10 F(I)=F(I)/10000
11 N(I)=LL(I)*F(I)
12 NEXT I
13 FOR I=1 TO 6
14 READ LLMEX(I)
15 LLMEX(I)=LLMEX(I)/1000
16 NEXT I
17 FOR I=1 TO 6
18 READ FMEX(I)
19 FMEX(I)=FMEX(I)/10000
20 NEXT I
21 LPRINT "IRISH MORTALITY AND IRISH FERTILITY"
22 GOSUB 43
23 DATA 4321,4232,4127,4011,3882,3740
24 DATA 23,347,805,897,660,324
25 DATA 4344,4296,4230,4156,4061,3955
26 DATA 492,1436,1545,1230,1015,472
27 FOR I=1 TO 6
28 N(I)=LL(I)*FMEX(I)
29 NEXT I
30 LPRINT "IRISH MORTALITY AND MEXICAN FERTILITY"
```

84   IV. Births and Deaths Under Stability

```
31 GOSUB 43
32 FOR I=1 TO 6
33 N(I)=LLMEX(I)*F(I)
34 NEXT I
35 LPRINT "MEXICAN MORTALITY AND IRISH FERTILITY"
36 GOSUB 43
37 FOR I=1 TO 6
38 N(I)=LLMEX(I)*FMEX(I)
39 NEXT I
40 LPRINT "MEXICAN MORTALITY AND MEXICAN FERTILITY"
41 GOSUB 43
42 END
43 R=.01
44 'THE LOOP IS DESIGNED TO CONTINUE UNTIL RSTAR-R=0
45 WHILE RSTAR-R
46 R=RSTAR
47 A=0
48 FOR I=1 TO 6
49 A=A+EXP((15-5*I)*R)*N(I)
50 NEXT I
51 RSTAR=LOG(A)/27.5
52 WEND
53 LPRINT USING "#####.#####";RSTAR
54 RETURN


IRISH MORTALITY AND IRISH FERTILITY
     0.00642
IRISH MORTALITY AND MEXICAN FERTILITY
     0.03272
MEXICAN MORTALITY AND IRISH FERTILITY
     0.00748
MEXICAN MORTALITY AND MEXICAN FERTILITY
     0.03364
```

The effects of mortality and fertility are easily found:

Fertility: $0.03272 - 0.00642 = 0.02630$ at Irish mortality level.

Mortality: $0.03364 - 0.03272 = 0.00092$ at Mexican fertility level.

Interaction: $0.00642 - 0.03272 - 0.00748 + 0.03364 = -0.00014$.

The interaction being 0.00014, and small in comparison with the main effects, we conclude that fertility is about 25 times as important as mortality in explaining the difference in rate of increase between Ireland in 1926 and Mexico in 1962.

6. The derivative of the logarithm in Solution 3 is equal to the average departure of the age of childbearing from $27\frac{1}{2}$, which is usually 2 or 3 years. Dividing this by $27\frac{1}{2}$ gives a number of the order of 0.1, so we gain about one decimal place per iteration.

7. Since r is not the variable of integration in Solution 2 we can differentiate under the integral sign to find

$$\psi'(r) = - \int xe^{-rx}l(x)m(x)\,dx,$$

$$\psi''(r) = \int x^2 e^{-rx}l(x)m(x)\,dx.$$

Since both integrands are positive, $\psi'(r) < 0$ and $\psi''(r) > 0$.

8.

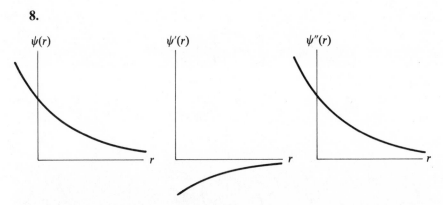

$\psi(r)$ $\psi'(r)$ $\psi''(r)$

9. The equation may be written $\psi(r) = 1$. From Solution 7 $\psi'(r)$ is negative for all r, which is to say that $\psi(r)$ is always decreasing. Such a function can pass only once through any real number, and in particular it can pass only once through unity. Hence $\psi(r) = 1$ can have only one real root.

10. If the age-specific birth rate for women aged x to $x + dx$ is $m(x)$, then the proportion of births to women 30 and over is

$$_{\beta-30}B_{30} = \frac{\int_{30}^{\beta} e^{-rx}l(x)m(x)\,dx}{\int_{\alpha}^{\beta} e^{-rx}l(x)m(x)\,dx},$$

where α is the lowest age of reproduction and β is the highest. The denominator here is equal to unity by virtue of the definition of r, the intrinsic rate of natural increase. All the formulas of the preceding questions apply as long as we replace $l(x)$ by $l(x)m(x)$, known as the net maternity function, so none of the algebra needs to be done over again. Since the range of ages of childbearing is small compared with that of the population as a whole the declining exponential of equation (2), in Solution 29, Ch. III, will now have a much less steep slope. For instance the mean age of those women bearing children after 30 might be 33, and the mean age of all women bearing children might be 25; the principal factor in the expression corresponding to (2) will be e^{-8r}.

The result of Question 29, Ch. III, is more general yet. If $m(x)$ be interpreted as the proportion of the population aged x to $x + dx$ in the labor force, and α and β the lower and upper limits of the range of working ages, then all the results apply once again.

11. Dividing by

$$R_0 = \int_{\alpha}^{\beta} l(x)m(x)\,dx,$$

taking logarithms and expanding the exponential under the integral sign to the term in r^2 gives

$$\ln \int_{\alpha}^{\beta} \left(1 - rx + \frac{r^2x^2}{2}\right) \frac{l(x)m(x)\,dx}{R_0} = -\ln R_0,$$

or

$$\ln\left(1 - \frac{rR_1}{R_0} + \frac{r^2\,R_2}{2\,R_0}\right) = -\ln R_0,$$

where

$$R_i = \int_\alpha^\beta x^i l(x)m(x)\,dx,$$

or by Taylor's theorem applied on the left,

$$-\frac{rR_1}{R_0} + \frac{r^2\,R_2}{2\,R_0} - \left(\frac{R_1}{R_0}\right)^2 \frac{r^2}{2} = -\ln R_0,$$

so that

$$\frac{\sigma^2}{2}r^2 - mr + \ln R_0 = 0.$$

12. Distinguish the higher mortality population by a star. Then

$$\mu^*(x) = \mu(x) + \delta$$

so

$$l^*(x) = e^{-\delta x}l(x),$$

and we can write

$$\int_\alpha^\beta e^{-(r^*+\delta)x}l(x)m(x)\,dx = 1.$$

Compare with the unstarred equation for the population with lower death rates, as given in (2). Because of the uniqueness of the real root, $r^* + \delta$ must be the same as r, so that $r^* = r - \delta$. If age-specific death rates are the same, an excess δ in death rates at all ages corresponds to a rate of increase lower by δ. The crude birth rate and the age distribution are the same for the two populations; the crude death rate for the one of higher mortality is higher by exactly δ.

13. If death rates drop by 0.001 at every age, 0.001 of the population will survive who would not otherwise survive. The population will be greater in the ratio 1.001 than it would otherwise be at all ages. Hence the age distribution is unchanged.

14. If the intrinsic rate is lower by absolute amount δ we have as the equation for δ,

$$\int_\alpha^\beta e^{-(r-\delta)x}l(x)(1 - f)m(x)\,dx = 1,$$

and remembering that $e^{\delta x}$ may be approximated by $1 + \delta x$, and that the original equation is satisfied by r, we have after cancellation, ignoring the term in $f \delta \bar{x}$,

$$(1 - f)\delta \bar{x} - f \doteq 0,$$

so that $\delta = f/[(1 - f)\bar{x}]$. The intrinsic birth rate is diminished by the quantity $f/(1 - f)$ divided by the mean age of childbearing in the stable population.

15. We need to solve for the fraction f the equation

$$\int_{\alpha}^{\beta} e^{-rx} e^{\delta x} l(x)(1 - f)m(x)\, dx = 1.$$

As before the equation reduces to $(1 - f)\delta \bar{x} - f \doteq 0$, so $f = \delta \bar{x}/(1 + \delta \bar{x})$. That is, the birth rates at all ages are lower by $100 \delta \bar{x}/(1 + \delta \bar{x})$ percent.

16. At birth the probability of surviving to each age x is $l(x)$, and the chance of childbearing between age x and $x + dx$ is $m(x)\, dx$; the total of expected births is the integral

$$\int_{\alpha}^{\beta} l(x)m(x)\, dx = R_0, \quad \text{say.}$$

When $R_0 = 1$ the population just replaces itself and the equation in (2) is satisfied by $r = 0$.

17. The argument is similar to that used for a somewhat different purpose in the preceding questions. An increase equal to r in the force of mortality lowers the survivorship from $l(x)$ to $e^{-rx} l(x)$; entering this last in place of $l(x)$ in R_0 of solution (16) makes R_0 equal to unity, according to (2).

18.

$$\ln\left(\int_{-\infty}^{\infty} e^{rx} f(x)\, dx\right) = \ln\left(1 + r\bar{x} + \frac{r^2}{2!}\mu_2' + \cdots\right),$$

and expanding the logarithm gives

$$\bar{x}r + (\mu_2' - \bar{x}^2)\frac{r^2}{2!} + \cdots,$$

so κ_1 may be identified with $\bar{x}$ and κ_2 with $\mu_2' - \bar{x}^2$, i.e., with the variance.

19. If we divide both sides of the characteristic equation in (2) by R_0, take logarithms, and change signs, we have on the right the cumulant-generating function of the net maternity function, and on the left the logarithm of R_0:

$$\ln R_0 = \kappa_1 r - \frac{\kappa_2 r^2}{2!} + \frac{\kappa_3 r^3}{3!} + \cdots,$$

where κ_i is the ith cumulant of the net maternity function. Hence $T \doteq \ln R_0/r$ is given by this series divided by r:

$$T = \kappa_1 - \frac{\kappa_2 r}{2!} + \frac{\kappa_3 r^2}{3!} + \cdots.$$

20. The mean age of childbearing in the stationary population is κ_1. The mean age of childbearing in the stable population is

$$\int_\alpha^\beta x e^{-rx} l(x) m(x) \, dx,$$

and the logarithm of this is the cumulant-generating function of $xl(x)m(x)$. It can be shown that this series starts with $\kappa_1 - \kappa_2 r$. The average of this and κ_1 is $\kappa_1 - (\kappa_2 r)/2!$, the first two terms in the expression for T above.

21. Most populations show a length of generation for females between 25 and 33 years. (In some 800 populations for which the calculation was made only about 1 percent fell outside this range.) Using $r = \ln R_0/T$ shows that r must be above $\ln R_0/33$ and below $\ln R_0/25$, i.e.,

$$0.03 \ln R_0 < r < 0.04 \ln R_0.$$

22. To find this one has to solve two equations and take the difference of their results. One equation is

$$\int_\alpha^{35} e^{-rx} l(x) m(x) \, dx + \frac{3}{4} \int_{35}^\beta e^{-rx} l(x) m(x) \, dx = 1,$$

which will give a value of r somewhat lower than the root of

$$\int_\alpha^\beta e^{-rx} l(x) m(x) \, dx = 1.$$

A first approximation to the solution is $r - g/4\kappa_1$, where g is the fraction of fertility that occurs over age 35.

23. Using the exact relation $r = \ln R_0/T$ makes the result obvious. It is also obvious from a plot of the expression

$$\psi(r) = \int_\alpha^\beta e^{-rx} l(x) m(x) \, dx$$

as a function of r; R_0 appears as the intercept $\psi(0)$ on the ordinate, and r is the abscissa of the intersection of $\psi(r)$ with the horizontal line unit distance above the origin.

24. The characteristic equation in this special case is

$$e^{-23r} l(23) + e^{-30r} l(30) = 1.$$

To solve multiply both sides by e^{25r}, take logarithms and divide by 25. The resultant iterative formula is

$$r^* = \tfrac{1}{25} \ln(0.95e^{2r} + 0.93e^{-5r}),$$

and the successive iterates are 0, 0.02525, 0.02393, 0.02399, the last being correct to five places.

25. Dividing the characteristic equation by R_0 and taking the logarithm generates the cumulants of the net maternity function on the right, so we have as the equivalent to the equation (2):

$$r\kappa_1 - \frac{r^2}{2!}\kappa_2 + \frac{r^3}{3!}\kappa_3 + \cdots = \ln R_0.$$

If all cumulants beyond the second are zero we have

$$\frac{r^2}{2!}\kappa_2 - r\kappa_1 + \ln R_0 = 0.$$

This is the same result as would be obtained by fitting a normal curve to the net maternity function $l(x)m(x)$ in (2).

26. Call the expression on the left in Solution 25 $f(r)$. The function $f(r)$ will help find the derivative of r with respect to the mean κ_1, using the theory of implicit functions. The total differential of the function is

$$df = \frac{\partial f}{\partial r}dr + \frac{\partial f}{\partial \kappa_1}d\kappa_1 = 0,$$

in which we can solve for the differential $d\kappa_1/dr$ as a ratio of partials:

$$\frac{dr}{d\kappa_1} = -\frac{\partial f}{\partial \kappa_1} \bigg/ \frac{\partial f}{\partial r},$$

or in the particular case of Solution 25

$$\frac{dr}{d\kappa_1} = -(-r)/(r\kappa_2 - \kappa_1) = \frac{r}{r\kappa_2 - \kappa_1}.$$

A decrease of 1 year in κ_1 increases r by $r/(r\kappa_2 - \kappa_1)$. If $\kappa_2 = 40$ and κ_1 is 27, then we have

r	$\dfrac{r}{r\kappa_2 - \kappa_1} = \dfrac{r}{40r - 27}$
0.00	0.00000
0.01	-0.00038
0.02	-0.00076
0.03	-0.00116
0.04	-0.00157

produced by the very simple program:

```
10 FOR R = 0 TO .04 STEP .01
20 CHANGE = R/(40*R-27)
30 LPRINT USING "###.##";R;
40 LPRINT USING "#####.#####";CHANGE
50 NEXT R
```

```
0.00     0.00000
0.01    -0.00038
0.02    -0.00076
0.03    -0.00116
0.04    -0.00157
```

Evidently age of childbearing makes no difference if the rate of increase is zero, and even with the most rapid possible rate of increase, having children a year younger will only raise the rate by about $1\frac{1}{2}$ points per thousand population.

27. A functional iteration for r is

$$r^* = \frac{\ln R_0}{\kappa_1 - \kappa_2 r/2}.$$

Solving this for $\kappa_1 = 26\frac{1}{2}$ and $27\frac{1}{2}$, with $\kappa_2 = 40$ and various values of R_0 gives the answer. For instance, with $R_0 = 2.3$ we have after one iteration,

$$\text{with } \kappa_1 = 26\frac{1}{2}, \quad r = 0.03143,$$

$$\text{with } \kappa_1 = 27\frac{1}{2}, \quad r = 0.03029,$$

a difference of 0.00114, against the -0.00116 shown above. For complicated derivatives this technique of arithmetical checking of algebra is indispensable.

28. $1/(\mu + r)$ is the mean age at death in the increasing population and $1/\mu$ is the age at which an individual expects to die. The difference can be large. If $r = 0.03$ and $\mu = 0.015$, then we have $1/(0.03 + 0.015) = 22.2$ years in the population and $1/0.015 = 66.7$ years for the individual.

29. In a stable population the number of persons at age a, per one of the total population, is $be^{-ra}l(a)$, so the integral of this quantity over all ages must be unity, and hence

$$b = \frac{1}{\int_0^\infty e^{-ra}l(a)\,da}.$$

The population whose death rates are higher by δ at all ages will have

$$l^*(a) = \exp\left(\int_0^a -(\mu(x) + \delta)\,dx\right) = e^{-\delta a}l(a),$$

and its r^* was stated to be $r - \delta$. So we have for the birth rate of the higher-mortality population

$$b^* = \frac{1}{\int_0^\infty e^{-r^*a}l^*(a)\,da} = \frac{1}{\int_0^\infty e^{-(r-\delta)a}e^{-\delta a}l(a)\,da}$$

$$= \frac{1}{\int_0^\infty e^{-ra}l(a)\,da} = b.$$

Such a result is possible because the special kind of difference in mortality (δ, the same quantity for all ages) does not affect the age distribution. If two populations have the same age distribution and the same age-specific birth rates, then they have the same crude or overall birth rate.

30. The problem is to find r^* given

$$b^* = \frac{1}{\int_0^\infty e^{-r^*a}e^{-\delta a}l(a)\,da}$$

$$= \frac{1}{\int_0^\infty e^{-(r^*+\delta)a}l(a)\,da} = b.$$

Since b is the reciprocal of $\int_0^\infty e^{-ra}l(a)\,da$ this shows that

$$\int_0^\infty e^{-(r^*+\delta)a}l(a)\,da = \int_0^\infty e^{-ra}l(a)\,da.$$

But because $\int_0^\infty e^{-ra}l(a)\,da$ is a monotonic function of r it can have only one root in r when equated to a constant. From this uniqueness it follows that the two integrals can only be equal if $r^* + \delta = r$, and hence $r^* = r - \delta$.

31. Call the initial common rate of increase r, the two survivorships $l^*(a)$ and $l(a)$, the two birth rates b^* and b. Then the initial crude death rate is

$$(cdr)_0 = ((b^* - r) + (b - r))/2.$$

If the procedure lowers mortality by ε uniformly at all ages for the starred subpopulation it will increase its r to $r + \varepsilon$, and will have no effect on its crude birth rate b^*. The population that results at time t will have a weight of $e^{\varepsilon t}$ for the starred subpopulation and unity for the other, so its crude death rate will be

$$(cdr)_t = (e^{\varepsilon t}(b^* - r - \varepsilon) + (b - r))/(e^{\varepsilon t} + 1).$$

Then $(cdr)_t$ will be greater than $(cdr)_0$ if and only if

$$\frac{e^{\varepsilon t}(b^* - r - \varepsilon) + (b - r)}{e^{\varepsilon t} + 1} > \frac{b^* - r + (b - r)}{2}.$$

Ultimately as t becomes large, the condition becomes $\varepsilon < (b^* - b)/2$ which is the same as

$$\varepsilon < \frac{(b^* - r) - (b - r)}{2}.$$

Expressed in words, the result is that sooner or later the death rate will rise as a result of the improvement in mortality if the improvement is of amount less than half of the initial difference in death rates. Strictly speaking, these statements should be expressed in the language of comparative statics, but that would make the exposition tedious.

32. The numerator of the derivative $s'(t)$ is equal to

$$(\omega(t) + 1)(\omega'(t)a) - (\omega(t)a + 1)(\omega'(t)) = \omega'(t)(a - 1) < 0,$$

since $a < 1$, and the denominator, a square, is positive. Hence $s'(t)$ is negative, so $s(t)$ must be a decreasing function of t.

CHAPTER V
Projection and Forecasting

Statistical data regarding population, like all other data, refer to the past; action and policy require knowledge of the future. What counts is not what the population has been, but what it will be. The cost of an annuity taken out today depends on future not on past mortality. Whatever continuities exist between past and future must be discovered and exploited if decisions regarding the future are to be soundly based, from the price of an annuity to the location of a retail store.

The questions in this chapter concern continuities between past and future. Among other matters we search for functions of past observations that are invariant with time, starting with the simplest cases. Techniques are most often illustrated with fixed parameters, using models that in other respects as well are less elaborate than those applied in professional practice.

5.1. Problems

5.1.1. Extrapolation of Population Totals

1. A curve of the form $P_t = P_0 e^{rt}$ has constant ratios in successive periods; one of the form $P_t = P_0 + at$ has constant differences. Prove these statements.

2. What function of the population at successive moments of time is constant in the case of the hyperbola $P_t = P_0/(t_e - t)$?

3. What function of the population at successive moments is constant for the logistic $P_t = a/(1 + e^{-r(t - t_0)})$?

4. If P_1, P_2, and P_3 are the populations of an area at three equidistant time periods, prove that the ultimate population α on the logistic through them is given by

$$\alpha = \frac{\dfrac{1}{P_1} + \dfrac{1}{P_3} - \dfrac{2}{P_2}}{\dfrac{1}{P_1 P_3} - \dfrac{1}{P_2^2}}.$$

where numerator and denominator are both supposed positive.

5. For the U.S. census of 1810 the population was approximately 7290; that of 1860, 30,443; that of 1910, 91,661, all in thousands. Find the asymptote α from the preceding formula.

6. Prove that the nth finite difference of a curve of the nth degree is constant.

7. Following are the counts of population in the United States at successive censuses:

1940	132	1960	179
1950	151	1970	203

Difference the table, and calculate the value for 1980 by linear, quadratic, and cubic polynomials, using successive differences. Investigate the accuracy of these several methods.

8. Continue the experiment with the birth series for the United States by calendar years from 1962 to 1978 (*Statistical Abstracts*, 1980, p. 61). Try cubic, quadratic, and linear projections, as well as assuming no change from the preceding 5-year amount. Compare the accuracy of these several "methods."

9. Using only the expression of Problem 4, and supposing a population to be counted (in thousands) for

1940 at 139,288

1950 at 156,102

1960 at 171,966

calculate the ultimate population on the logistic, α of Problem 4, and also the population of 1970, 1980, 1990, and 2000.

10. The populations at time $t - 2$, $t - 1$, and t are each independently estimated with variance σ^2. What is the variance of the projection to time $t + n$ if the fitting is: (a) a straight line; (b) a quadratic?

11. For the U.S. series of censuses from 1790 to 1980, as given in the current *Statistical Abstract*, work out estimates that can be compared census by census, based on:

(1) Accepting the previous census figure for this census—S (for same).
(2) Arithmetic progression—A.
(3) Quadratic extrapolation—Q.
(4) Cubic extrapolation—C.
(5) Geometric progression—G.
(6) Logistic fit—L.
(7) Hyperbola or harmonic progression—H.

12. Fibonacci described a rabbit population such that each pair bears a pair when it is 1 month old and another pair when it is 2 months old. Starting with one pair, this gives rise to the series 1, 1, 2, 3, 5, 8, ... State the general term of the series and find its ultimate stable rate of increase. How must the model be elaborated if it is to apply to a human population?

13. One would like to use the theory of the demographic transition to forecast population. Suppose a population with a high birth rate and a death rate equal to its birth rate, then a decline in the death rate followed by a decline in the birth rate, until the stationary condition resumes. Show that the increase over the transition is e^A, where A is the sum of the (crude) birth rates over the transition minus the sum of the death rates.

14. If the birth and death rates are identical decreasing curves, starting together and each falling by the amount K, with the death rates preceding the birth rates by L years, then the ratio of increase over the transition is e^{KL}. How does the increase depend on the lag L?

5.1.2. Matrix Technique

15. Draw up a matrix **S** that will project the part initially alive of one sex of a population. What power of the matrix consists wholly of zeros? Interpret this nilpotent property against what you know of the lifetimes of people already alive at the time of the projection.

16. What matrix will serve to project births and hence the youngest age group of the population?

5.1.3. Analysis of the Projection Matrix

17. The United States female projection matrix for ages 0–44 in 15-year age groups with 1965 age-specific birth and death rates is

$$\mathbf{M} = \begin{bmatrix} 0.4271 & 0.8498 & 0.1273 \\ 0.9924 & 0 & 0 \\ 0 & 0.9826 & 0 \end{bmatrix}.$$

The age distribution for females for 1965 is 0–14, 29,415; 15–29, 20,886; 30–44, 18,040; Total, 68,341, all in thousands. Project to 1980 and 1995.

18. Square the matrix in Problem 17 to find the projection matrix that will carry the population over 30 years. Premultiply the 1965 vector by this square matrix, and see whether you obtain the same result for 1995 as in Problem 17.

19. The right-hand stable vector $\mathbf{P}$ of the matrix $\mathbf{M}$ satisfies the equation $\mathbf{MP} = \lambda\mathbf{P}$, where λ is a suitable scalar. Express this condition in words.

20. Take the matrix in Problem 17 to its 32nd and 33rd powers, and find how fast the population is increasing per year after 32 periods of 15 years.

21. What is the characteristic equation for the U.S. projection matrix in Problem 17?

22. What is the positive root of the preceding equation?

23. Using the derivative $d\lambda^*/d\lambda$ find how fast the preceding functional iteration ought to converge in theory.

24. Compare the above functional iteration with that obtained by the Newton–Raphson procedure.

25. Prove that the matrix of Problem 17 satisfies its characteristic equation.

26. Following is the projection matrix in 15-year age groups for Norway females in 1964:

$$\mathbf{M} = \begin{bmatrix} 0.3476 & 0.8917 & 0.1577 \\ 0.9952 & 0 & 0 \\ 0 & 0.9909 & 0 \end{bmatrix}.$$

By projection of an arbitrary vector find the ultimate stable ratio of population growth that it implies, as well as the corresponding stable age distribution; set out the characteristic equation and solve for the three roots.

27. What is the net reproduction rate R_0 of the regime of mortality and fertility contained in the matrix of Problem 26?

28. What is the mean age of childbearing or length of generation T in the matrix of Problem 26?

29. How would one reduce to zero the rate of increase of the regime of Problem 26 without changing either the mortality rates or the relative fertility at different ages, i.e., by changing fertility at all ages in proportion?

30. Supposing that the matrix of Problem 26 holds for 1964 to 2009, and that the stationary matrix of Problem 29 applies in the year 2009 and subsequently, given that the initial distribution in the three age groups is 470, 387, 352 (total, 1209), track out the female population under age 45 until the end of the 21st century.

31. In a heterogeneous population, show that a forecast made for the whole population at its average initial rates gives a smaller result than the total of the projections of subpopulations each at its own initial rate of increase. Use the matrix of Problem 26, adding and subtracting 0.1 to each element of the first row to obtain the regimes for the subpopulations.

32. For the five life tables of the United States at successive decades, call the first the standard life table, say $l_s(x)$, transform to logits,

$$Y_x = \tfrac{1}{2} \ln\left(\frac{1 - l_x}{l_x}\right),$$

then fit the first, taken as a standard, to each of the last four tables, expressing their logits as

$$Y_x = \alpha_i + \beta_i Y_{sx}, \qquad i = 2, 3, 4, 5.$$

Examine the feasibility of projecting the α and β and so obtaining Y_x pertaining to the future from which survivorships can be calculated.

5.1.4. Breakdown by Sex and Other Attributes

33. Expand the matrix $\mathbf{M} = \mathbf{S} + \mathbf{B}$ to project the two sexes separately with mixed dominance, supposing s boys to each girl baby at each age of childbearing. Let the fraction D of each birth come from the mother and $1 - D$ from the father.

34. Sketch out the block matrix projecting the population according to rural and urban residence, age, and sex.

35. From the form of the female dominant matrix in Problem 33, show that the initial number of males, as well as their mortality rates, make no difference to the ultimate rate of increase of the population.

5.2. Solutions

1. If $P_t = P_0 e^{rt}$ and $P_{t+1} = P_0 e^{r(t+1)}$, then by division $P_{t+1}/P_t = e^r$, and this ratio of successive terms is not a function of time t, i.e., it is a constant in the meaning of the question. Similarly for $P_t = P_0 + at$ and $P_{t+1} = P_0 + a(t + 1)$; subtraction gives $P_{t+1} - P_t = a$, so the difference of successive terms is a constant.

2. The difference of the reciprocals of the tth and $(t + 1)$th terms is equal to $1/P_0$, which is independent of time.

3. The reciprocal $1/P_t$ is $(1 + e^{-r(t-t_0)})/a$, and subtracting this from the corresponding value for $1/P_{t+1}$ gives $e^{-r(t-t_0)}(e^{-r} - 1)/a$. The difference of reciprocals $1/P_{t+2} - 1/P_{t+1}$ is the same, but with $t + 1$ entered in place of t. Dividing the first difference by the second gives

$$\frac{\dfrac{1}{P_{t+1}} - \dfrac{1}{P_t}}{\dfrac{1}{P_{t+2}} - \dfrac{1}{P_{t+1}}} = e^r,$$

which is constant.

4. The problem is to eliminate the constants b and r among the equations

$$P_t = \frac{\alpha}{1 + be^{-rt}}, \qquad t = -1, 0, 1,$$

where the values of t are chosen for ease in algebraic manipulation. Taking reciprocals, and then eliminating b and r gives the required value. Note that by no means all triplets of points can be fitted with a real logistic.

5. 197 million.

6. If the curve is of the form $at^n + bt^{n-1} \ldots$, then the first difference is

$$a(t + 1)^n + b(t + 1)^{n-1} + \cdots - at^n - bt^{n-1} - \cdots = nat^{n-1} + \cdots$$

which is of degree lower by one than the original function. By successive differencing in this way it turns out that the nth difference is $n!\, a$, a constant.

7. Differencing out the table and continuing it down as though some order of differences is constant is the general approach. Projection with a cubic curve, i.e., supposing third differences constant, permits building up successive later values of the curve by adding: $-13 - 4 = -17$; $-17 + 24 = 7$; $7 + 203 = 210$, etc.

	Observed	Δ	Δ^2	Δ^3
1940	132			
		19		
1950	151		9	
		28		-13
1960	179		-4	
		24		-13
1970	203		-17	
	Calculated	7		
1980	210			

Similarly with a quadratic curve, i.e., supposing second differences constant at -4 gives 223 million, and with a straight line, i.e., supposing first differences constant at 24 gives 227 million.

Notice that the 1980 census count was 226.5 million, so the straight-line projection from 1960 and 1970 alone was more accurate than the quadratic or cubic using more data. Such an outcome is not unusual—the more elaborate the method the poorer the results unless the elaboration is appropriate.

Alternatively, if fourth differences are zero, in symbolic terms $\Delta^4 = 0$, and if E is the shift operator, we have by the binomial theorem,

$$\Delta^4 = (E - 1)^4 = E^4 - 4E^3 + 6E^2 - 4E + 1 = 0,$$

or in terms of our numbers, for a cubic fitting in millions,

$$P_{1980} = 4P_{1970} - 6P_{1960} + 4P_{1950} - P_{1940} = 210.$$

For a quadratic fitting,

$$P_{1980} = 3P_{1970} - 3P_{1960} + P_{1950} = 223,$$

and for a linear fitting,

$$P_{1980} = 2P_{1970} - P_{1960} = 227.$$

Calculation by the same three formulas for the series of 20 censuses since 1790 gives a mean absolute error in millions of

> 8.06   for the cubic,
>
> 4.00   for the quadratic,
>
> 2.89   for the straight line.

With geometric rather than arithmetic series, i.e., taking logarithms and subsequently exponentiating, we find

> 10.56   for the cubic,
>
> 5.35   for the quadratic,
>
> 3.16   for the straight line.

As an experiment on short-term forecasts, one can take the 20 published estimates of U.S. population at midyear 1957 to 1976. The mean errors, again in millions, are

	Arithmetic	Geometric
Cubic	0.369	0.377
Quadratic	0.218	0.221
Linear	0.139	0.156

As in the 10-year forecasts, the linear shows less than half the error of the cubic. The examples warn us against the gratuitous errors introduced by inappropriate elaboration.

8. We find, in thousands of births, the mean error:

	Arithmetic	Geometric
Cubic	240	226
Quadratic	156	148
Linear	114	112
No change from preceding year	100	100

The "no change" estimate is the best, followed by the linear. This time the geometric wins by a small margin over the arithmetic.

9. From the formula in Question 4 we have $\alpha = 256{,}427$, and

1970	186,446	1990	210,323
1980	199,269	2000	219,634

10. The arithmetic projection to time $t + n$ based on time $t - 1$ and t may be written as

$$P_{t+n} = P_t + n(P_t - P_{t-1}) = (n + 1)P_t - nP_{t-1},$$

$$\text{Var } P_{t+n} = (n + 1)^2 \text{ Var } P_t + n^2 \text{ Var } P_{t-1},$$

supposing independence of errors. Hence

$$\text{Var } P_{t+n} = (2n^2 + 2n + 1)\sigma^2.$$

For a quadratic we have as earlier:

$$P_{t+n} = E^{n+2}P_{t-2} = (1 + \Delta)^{n+2}P_{t-2}$$

$$= P_{t-2} + (n + 2)(P_{t-1} - P_{t-2})$$

$$+ \frac{(n + 2)(n + 1)}{2!}(P_t - 2P_{t-1} + P_{t-2})$$

$$= P_t\left(\frac{(n + 2)(n + 1)}{2}\right) - P_{t-1}((n + 2)(n)) + P_{t-2}\left(\frac{(n + 1)n}{2}\right),$$

$$\text{Var}(P_{t+n}) = \left(\frac{(n + 2)^2(n + 1)^2}{2^2} + (n + 2)^2 n^2 + \frac{(n + 1)^2 n^2}{2^2}\right)\sigma^2.$$

11. The table shows departures of the seven estimates from the census figure, and the standard error of these departures. Notice that the cubic is the second worst, the arithmetic is the best, the hyperbolic better than the more reasonable logistic.

Year				Departures			
	S	A	Q	C	G	L	H
1820	−2.40	−0.50	0.00	3.00	0.18	0.18	1.62
1830	−3.30	−0.90	−0.40	−0.40	−0.10	−0.39	1.50
1840	−4.20	−0.90	0.00	0.40	0.23	0.42	2.56
1850	−6.10	−1.90	−1.00	−1.00	−0.53	−0.92	2.16
1860	−8.20	−2.10	−0.20	0.80	0.08	1.19	4.67
1870	−8.40	−0.20	1.90	2.10	2.70	2.56	8.77
1880	−10.40	−2.00	−1.80	−3.70	0.25	−2.83	4.14
1890	−12.70	−2.30	−0.30	1.50	0.42	0.04	5.06
1900	−13.10	−0.40	1.90	2.20	2.81	2.18	8.20
1910	−16.00	−2.90	−2.50	−4.40	−0.17	−3.43	3.99
1920	−13.70	2.30	5.20	7.70	5.67	5.92	10.83
1930	−17.10	−3.40	−5.70	−10.90	−1.36	−6.43	1.39
1940	−8.90	8.20	11.60	17.30	10.97	12.99	14.80
1950	−19.60	−10.70	−18.90	−30.50	−10.05	−15.49	−9.31
1960	−28.00	−8.40	2.30	21.20	−5.48	26.83	−1.55
1970	−23.90	4.10	12.50	10.20	9.28	19.31	16.82
1980	−18.80	5.10	1.00	−11.50	8.29	0.01	12.45
R.M.S.	14.47	4.47	6.56	11.24	5.18	9.74	8.07

12. The general term is $B_t = B_{t-1} + B_{t-2}$, with time in months and B_t the number of pairs born at time t. Dividing by B_{t-1} gives

$$\frac{B_t}{B_{t-1}} = 1 + \frac{B_{t-2}}{B_{t-1}}.$$

At stability $B_t/B_{t-1} = B_{t-1}/B_{t-2} = R$, say. Then the equation is

$$R = 1 + \frac{1}{R} \quad \text{or} \quad R^2 - R - 1 = 0,$$

so $R = (1 \pm \sqrt{5})/2$. The relevant root is then 1.618.

Even easier than solving the quadratic is tracking out the curve by computer. Following is the program and output. Note that within eight cycles of projection the ratio is correct to three significant figures, and within ten is correct to four.

Extension to a human population requires only changing the unit from the month to the year or 5-year period and increasing the number of times of reproduction from 2 months to six or seven 5-year periods.

```
10 'TRACKING OUT THE FIBONACCI RABBITS
20 DIM B(100),R(100)
30 B(0)=0
40 B(1)=1
50 FOR I=1 TO 99
60 B(I+1)=B(I)+B(I-1)
70 R(I+1)=B(I+1)/B(I)
80 LPRINT I,B(I),R(I)
90 IF I<2 THEN GOTO 110
100 IF ABS(R(I)-R(I-1))<.0001 THEN STOP
110 NEXT I
```

I	$B(I)$	$R(I)$
1	1	0
2	1	1
3	2	2
4	3	1.5
5	5	1.66667
6	8	1.6
7	13	1.625
8	21	1.61538
9	34	1.61905
10	55	1.61765
11	89	1.61818
12	144	1.61798
13	233	1.61806

13. At any time the ratio of increase is $e^{b(t)}/e^{d(t)}$, where b and d are the crude birth and death rates, respectively, and are taken as functions of time. Then if the transition starts at time t_0 and ends at time t_1, we have for the ratio of increase over t_0 to t_1,

$$\text{Ratio} = \exp\left(\int_{t_0}^{t_1}(b(t) - d(t))\, dt\right) = e^A,$$

where A is as defined in the problem, and on a chart would be the area enclosed by the birth and death curves.

14. $de^{KL}/dL = Ke^{KL}$. The population ratio over the transition is directly proportional to K, the difference between initial and final birth (or death) rates, and to the exponential of the area enclosed by the birth and death curves.

15. The matrix is $n \times n$ in size, where $n = 18$ for a population classified into 5-year age groups and of which the last members die before age 90; it is 90×90 in size for 1-year age intervals. The only non-zero elements are in the subdiagonal, and the ith element of the $(i + 1)$th row is $_5L_{5xi}/_5L_{5x(i-1)}$, $i = 1, 2, \ldots$. The nth power of the matrix consists wholly of zeros, since the first power has zeros in its first row, the square has zeros in its first two rows, $\ldots$, the $(n - 1)$th power has zeros in its first $n - 1$ rows. The nth power of the matrix projects the initial population through n time periods, and if no one lives to be as much as n years old there can be no one living at time n.

16. The initial age distribution is $\mathbf{P}_0$, the projection after 5 years is $\mathbf{SP}_0$, and so the exposure to the risk of childbearing over the 5-year interval in the several ages is given by the vector

$$\tfrac{5}{2}(\mathbf{P}_0 + \mathbf{SP}_0).$$

Call the (horizontal) vector of age-specific fertility rates $\mathbf{F}$. Then the expected births over the first 5-year period are the scalar quantity,

$$\tfrac{5}{2}\mathbf{F}(\mathbf{P}_0 + \mathbf{SP}_0),$$

and the survivors among these to the end of the period are the same multiplied by $_5L_0/5$:

$$\tfrac{5}{2}\mathbf{F}(\mathbf{I} + \mathbf{S})\mathbf{P}_0 \, _5L_0/5.$$

The part of this that does not include $\mathbf{P}_0$ is a horizontal vector that can be put into the first row of a matrix that has zeros in all other rows. If this matrix is called $\mathbf{B}$, then the complete projection matrix is $\mathbf{S} + \mathbf{B}$, where $\mathbf{S}$ has elements $_5L_{5xi}/_5L_{5x(i-1)}$ in its ith row, and $\mathbf{B}$'s first row is the horizontal vector $_5L_0 \mathbf{F}(\mathbf{I} + \mathbf{S})/2$. We will write $\mathbf{S} + \mathbf{B} = \mathbf{M}$.

17. Multiplying to form $\mathbf{MP}_0$ and $\mathbf{M}^2\mathbf{P}_0$ gives

Age	1980	1995
0–14	32,609	41,346
15–29	29,191	32,361
30–44	20,523	28,683

The following program produces population projections and powers of matrices. Its current output shows projections for 1995 (15) 2055, and the first four powers of $\mathbf{M}$. The program could be used in problems 17, 20, and 25.

```
10 'FILE MATEMULT
20 ' POPULATION PROJECTION AND
30 ' MULTIPLICATION OF MATRICES
40 'TO USE ENTER THE SIZE OF THE SQUARE MATRIX AS N
50 'ENTER THE NUMBER OF CYCLES AS E
60 'ENTER THE MATRIX AS DATA
70 'ENTER THE VECTOR IF THERE IS ONE AS DATA
80 N=3:E=5 'SIZE OF MATRIX AND NUMBER OF CYCLES OF PROJECTION
90 GOSUB 130 'FOR POPULATION PROJECTION
100 STOP
110 GOSUB 430 'FOR SUCCESSIVE POWERS OF MATRIX
120 STOP
130 DIM M(N,N),MM(N,N)
140 DIM PROD(N,N)
150 D=1
160 FOR I=1 TO N
```

```
170 FOR J=1 TO N
180 READ M(I,J)
190 MM(I,J)=M(I,J)
200 NEXT J
210 NEXT I
220 FOR I=1 TO N
230 READ P(I)
240 NEXT I
250 DATA .4271,.8498,.1273
260 DATA .9924,0,0
270 DATA 0,.9826,0
280 DATA 29415,20886,18040
290 FOR I=1 TO N
300 V(I)=0
310 FOR K=1 TO N
320 V(I)=V(I)+M(I,K)*P(K)
330 NEXT K
340 PRINT V(I);
350 NEXT I
360 PRINT
370 FOR I=1 TO N
380 P(I)=V(I)
390 NEXT I
400 C=C+1
410 IF C<E THEN GOTO 290
420 RETURN
430 PRINT
440 PRINT "FOLLOWING IS POWER";D;"OF THE MATRIX"
450 FOR I=1 TO N
460 FOR J=1 TO N
470 PROD(I,J)=0
480 FOR K=1 TO N
490 PROD(I,J)=PROD(I,J)+M(I,K)*MM(K,J)
500 NEXT K
510 PRINT MM(I,J);
520 NEXT J
530 PRINT
540 NEXT I
550 FOR J=1 TO N
560 FOR I=1 TO N
570 MM(I,J)=PROD(I,J)
580 NEXT I,J
590 D=D+1
600 IF D<E THEN GOTO 430
601 END
```

```
 32608.6   29191.4   20522.6
 41346.5   32360.7   28683.5
 48810.7   41032.3   31797.7
 59764.1   48439.7   40318.3
 71821.8   59309.9   47596.9
```

```
FOLLOWING IS POWER 1 OF THE MATRIX
 .4271   .8498   .1273
 .9924   0   0
 0   .9826   0


FOLLOWING IS POWER 2 OF THE MATRIX
 1.02576   .488035   .0543698
 .423854   .843341   .126333
 .975132   0   0


FOLLOWING IS POWER 3 OF THE MATRIX
 .922426   .925111   .130579
 1.01796   .484325   .0539566
 .416479   .828667   .124134


FOLLOWING IS POWER 4 OF THE MATRIX
 1.31205   .912184   .117425
 .915416   .91808   .129586
 1.00025   .475898   .0530178
```

18. The associative law holds for matrices, so $\mathbf{M}^2\mathbf{P}_0$ is the same as $\mathbf{M}(\mathbf{M}\mathbf{P}_0)$, and $\mathbf{M}^2\mathbf{P}_0$ does in fact produce the numbers in Solution 17.

19. A stable age distribution is such that on projection it is preserved to within a constant factor.

20. Squaring five times to obtain $\mathbf{M}^{32}$, then multiplying by $\mathbf{M}$ gives $\mathbf{M}^{33}$. The ratio of any element of $\mathbf{M}^{33}$ to the corresponding element of $\mathbf{M}^{32}$ is an approximation to the 15-year ratio of increase λ. It turns out that under such constant rates the population ultimately increases in the ratio $\lambda = 1.2093$, in total and at every separate age, each 15 years. This is the stable or intrinsic ratio of increase.

21. The condition for consistency in the set of homogeneous equations $\mathbf{MP} = \lambda\mathbf{P}$ is that λ take a certain value, that which satisfies the characteristic equation. The characteristic equation in λ is the determinant $|\mathbf{M} - \lambda\mathbf{I}|$ equated to zero. Expanding by the rules for determinants we have

$$\lambda^3 - 0.4271\lambda^2 - 0.8433\lambda - 0.1241 = 0.$$

22. Separating λ^3, dividing by λ, then taking the square root, gives the functional iteration

$$\lambda^* = \sqrt{0.4271\lambda + 0.8433 + 0.1241/\lambda}.$$

Starting with $\lambda = 1$ gives iterates 1.1809, 1.2053, 1.2087, 1.2092.

23. By numerical differentiation on a computer we find $d\lambda^*/d\lambda = 0.06$ very nearly.

24. Newton–Raphson iteration in the equation $f(\lambda) = 0$ is based on a Taylor expansion to the linear term:

$$f(\lambda^*) = f(\lambda) + (\lambda^* - \lambda)f'(\lambda) = 0.$$

Solve for λ^* to obtain

$$\lambda^* = \lambda - \frac{f(\lambda)}{f'(\lambda)}.$$

In our case

$$\lambda^* = \lambda - \frac{\lambda^3 - 0.4271\lambda^2 - 0.8433\lambda - 0.1241}{3\lambda^2 - 0.8542\lambda - 0.8433}.$$

Iterates with the matrix of Problem 17 are 1, 1.3076, 1.2189, 1.2093. Convergence is slightly more rapid than in the method of Problem 22. (See Kellison (1975).)

25. The square matrix is

$$\mathbf{M}^2 = \begin{bmatrix} 1.0258 & 0.4880 & 0.0544 \\ 0.4239 & 0.8433 & 0.1263 \\ 0.9751 & 0 & 0 \end{bmatrix}$$

and the cube is

$$\mathbf{M}^3 = \begin{bmatrix} 0.9224 & 0.9251 & 0.1306 \\ 1.0180 & 0.4843 & 0.0540 \\ 0.4165 & 0.8287 & 0.1241 \end{bmatrix}.$$

Any corresponding set of elements satisfies the characteristic equation. For instance, the first elements of the first row entered in the equation give

$$0.9224 - (0.4271)(1.0258) - (0.8433)(0.4271) - 0.1241 = 0.$$

26. The stable ratio is 1.1973; the intrinsic rate is $\frac{1}{15}$ of the logarithm of this or 0.0120; the stable age distribution to within an arbitrary constant is 23.25 for age 0–14; 19.33 for age 15–29; 16.00 for age 30–44. The characteristic equation is $\lambda^3 - 0.3476\lambda^2 - 0.8874\lambda - 0.1555 = 0$, and its three roots are 1.1973, -0.6499, and -0.1998.

27. The expected number of girl children to a girl just born is the sum of the coefficients of the last three terms of the characteristic equation of Solution 26 with sign reversed, i.e., $R_0 = 0.3476 + 0.8874 + 0.1555 = 1.3905$.

28. If T is defined by $e^{rT} = R_0$, then

$$T = \frac{\ln R_0}{r} = \frac{\ln(1.3905)}{0.0120} = 27.47 \text{ years.}$$

29. By dividing each element of the first row by the net reproduction rate $R_0 = 1.3905$.

30.

```
10 'PROGRAM FOR PROJECTING POPULATION IN
20 'THREE AGE GROUPS, STARTING WITH THE REGIME OF NORWAY
30 'FEMALES FOR 1964, THEN CONTINUING FROM 1994 WITH
40 'DROP TO REPLACEMENT.
50 'FIRST READ IN THE MATRIX AND VERTICAL VECTOR
60 FOR I=1 TO 3
70 FOR J=1 TO 3
80 READ M(I,J)
90 NEXT J
100 NEXT I
110 DATA .3476,.8917,.1577,.9952,,,,,.9909,
120 P(1)=470:P(2)=387:P(3)=352
130 LPRINT 15*C+1979,
140 'THEN PREMULTIPLY THE AGE VECTOR BY THE MATRIX
150 FOR I=1 TO 3
160 FOR J=1 TO 3
170 P1(I)=P1(I)+M(I,J)*P(J)
180 NEXT J
190 'PRINT OUT THE RESULT
200 LPRINT USING "#########,###";P1(I),
210 T=T+P1(I)
220 NEXT I
230 LPRINT USING "#########,###";T
240 FOR I=1 TO 3
250 P(I)=P1(I)
260 P1(I)=0
270 NEXT I
280 T=0
290 C=C+1
```

```
300 IF C<3 THEN GOTO 130
310 'NOW LOWER THE MATRIX THAT IS TO BE APPLIED
320 'AFTER 1994 WHEN C=2
330 FOR J=1 TO 3
340 IF C=3 THEN M(1,J)=M(1,J)/1.3905
350 NEXT J
360 IF C<9  THEN GOTO 130
370 'THE REQUIRED OUTPUT FOLLOWS:
```

1979	564	468	383	1,415
1994	674	561	463	1,698
2009	808	670	556	2,034
2024	695	804	664	2,163
2039	765	692	797	2,253
2054	725	761	685	2,171
2069	747	721	754	2,222
2084	735	743	715	2,193
2099	741	731	737	2,209

31. We take it that one population is heterogeneous in that half has fertility factors 0.1 less and the other 0.1 more than the average of Norway 1964. If the population is projected to 2084 with the original matrix the result, as shown in the table below, is 5003 thousand. The alternative of projecting with lower fertility gives 1856 by 2084; with higher fertility, 11,353. If half the population grows with the low and half with the high we would have 1856/2 + 11,353/2 = 6,604, much more than the 5003 of the homogeneous projection.

	-15	15–29	30–44	Total
1979	564	468	383	1,415
1994	674	561	463	1,698
2009	808	670	556	2,034
2024	966	804	664	2,434
2039	1,157	962	797	2,916
2054	1,385	1,152	953	3,490
2069	1,659	1,379	1,141	4,179
2084	1,986	1,651	1,366	5,003
-0.1				
1979	443	468	383	1,294
1994	502	441	463	1,407
2009	500	500	437	1,437
2024	545	498	495	1,538
2039	558	542	493	1,593
2054	596	555	537	1,688
2069	618	593	550	1,760
2084	654	615	587	1,856

[*continued*

continued

	− 15	15–29	30–44	Total
+0.1				
1979	685	468	383	1,536
1994	869	682	463	2,014
2009	1,184	865	675	2,725
2024	1,562	1,179	857	3,598
2039	2,089	1,555	1,168	4,812
2054	2,778	2,079	1,540	6,397
2069	3,702	2,764	2,060	8,527
2084	4,929	3,684	2,739	11,353
	Average			
1979	1,415	1,294	1,536	1,415
1994	1,698	1,406	2,014	1,710
2009	2,034	1,436	2,724	2,080
2024	2,434	1,537	3,598	2,567
2039	2,915	1,592	4,811	3,202
2054	3,490	1,687	6,397	4,042
2069	4,179	1,760	8,526	5,143
2084	5,003	1,856	11,352	6,604

32. Reference: (Brass 1974). We leave this as a research question for the interested reader.

33. For a population with 18 5-year age intervals the projection matrix will again have 36 rows and 36 columns. If females are placed first in the matrix and the corresponding vector, and the span of fertile ages is 15–45, we will have for the first row D times the female fertility rates, averaged as in Problem 16 between successive ages. These will occupy the first 10 positions and later positions up to the 18th will be zero. For the 19th row we will have the same but multiplied by s. The subdiagonal from the second row to the 18th will be $_5L_{5x(i-2)}/_5L_{5x(i-3)}$, $i = 2$ to 18. For the 19th to 36th positions of both the first and the 19th rows the age-specific rates of fatherhood will be used, all multiplied by $1 - D$.

The projection matrix in block form, for female dominance, and arranged with females first, is

$$A = \begin{bmatrix} B + S & O \\ sB & S^* \end{bmatrix},$$

where each of the blocks is 18×18 for 5-year age groups to age 90, S^* is the survivorship matrix for males, O is an 18×18 matrix of zeros, and sB is the female birth matrix multiplied by the scalar s. Similarly for male dominance we have

$$A^* = \begin{bmatrix} S & B^*/s \\ O & B^* + S^* \end{bmatrix},$$

where **B*** is the matrix whose top row is age-specific fertility rates of men. For mixed dominance we take the weighted average:

$$\mathbf{A}D + \mathbf{A}^*(1 - D) = \begin{bmatrix} \mathbf{D}\mathbf{B} + \mathbf{S} & (1 - D)\mathbf{B}^*/s \\ s\mathbf{D}\mathbf{B} & \mathbf{B}^*(1 - D) + \mathbf{S}^* \end{bmatrix}.$$

34. The simplest representation is exactly the same as the matrix of Problem 33, but now in place of each element is a 2×2 block. Such blocks, both those for survivors and those for births, are broken down according to rural and urban, with the off-diagonal elements being migration.

35. Successive premultiplication by matrix **A** of Question 33 multiplies the males by a zero matrix in the second block of the first row, and by a survivorship matrix all of whose elements are less than unity in the second block of the second row. The square of the matrix must also have a zero block in its upper right-hand quarter and similarly for every higher power. The lower right of the powers of the matrix will be S^*, S^{*2}, etc., and so diminishing towards zero, since the female-dominant **A** has zeros only in its first row.

CHAPTER VI

Stochastic Population Models

In each person's life, certain events occur randomly. The sexes of one's future children are chance events. Probabilistic occurrences affect the size of one's family. The sizes of future generations from a specific couple are random variables. We will investigate the problem of finding the probability of extinction of a male line of descent. After a brief introduction to the needed mathematical concepts, we will study problems of the following type: If a newly married couple decided to have children until they have one of each sex, what is an estimate of the size of their completed family?

Pressures of overpopulation or of the fall in births have forced countries to look at the national effects of various family sizes. Although this book has presented many deterministic projection models for such problems, there are stochastic models which also help in population studies. The two types of models can serve to complement each other and provide fuller insights into population dynamics.

6.1. Needed Concepts

The sample space of an experiment is the set of all of its possible outcomes. It is denoted by S. Any subset E of S is called an event. If two events E and F have no points in common, i.e. $E \cap F = \varnothing$, E and F are said to be mutually exclusive. If $E_1, E_2, \ldots$ are events, then their union $\bigcup_{n=1}^{\infty} E_n$ is an event composed of all sample points which belong to at least one E_n. For any $E \subseteq S$, its complement E^c consists of points in S which are not in E. A family

$\mathcal{F}$ of events is a collection of subsets of S with the following properties:

(1) $S \in \mathcal{F}$
(2) If $E \in \mathcal{F}$, then $E^c \in \mathcal{F}$.
(3) If $E_1, E_2, \ldots$ belong to $\mathcal{F}$, then $\bigcup_{n=1}^{\infty} E_n \in \mathcal{F}$.

For each S and $\mathcal{F}$, we assume that a probability function $P(\cdot)$ has been defined on $\mathcal{F}$ with the properties:

(1) $P(E) \geq 0$ for $E \in \mathcal{F}$.
(2) $P(S) = 1$.
(3) If $E_1, E_2, \ldots$ belong to $\mathcal{F}$, and $E_j \cap E_k = \varnothing$ for $j \neq k$, then $P[\bigcup_{n=1}^{\infty} E_n] = \sum_{n=1}^{\infty} P(E_n)$.

If E and F are two events in S, with $P(F) > 0$, then the conditional probability that E occurs given that F has occurred is denoted and defined by

$$P(E|F) = \frac{P(E \cap F)}{P(F)}.$$

Two events E and F are called independent events if

$$P(E \cap F) = P(E)P(F).$$

Utilizing the conditional probability definition, this implies that $P(E|F) = P(E)$, i.e., knowing that F has occurred does not affect one's prediction of E occurring.

A random variable X is a real finite valued function defined on a sample space. The distribution function $F(\cdot)$ of the random variable X is defined for any real number x, $-\infty < x < \infty$, by

$$F(x) = P(X \leq x).$$

$F(x)$ has the properties:

(1) $F(x)$ is nondecreasing, $-\infty < x < \infty$;
(2) $F(x+) = F(x)$, i.e., $F(x)$ is continuous from the right for $-\infty < x < \infty$;
(3) $\lim_{x \to -\infty} F(x) = 0$;
(4) $\lim_{x \to +\infty} F(x) = 1$.

A random variable which can assume at most a countable number of possible values is called discrete, whereas one which assumes a continuum of possible values is called continuous. If X is a discrete random variable with values $x_1, x_2, \ldots$ then its probability function $p(x)$ is defined by

$$p(x) = P\{X = x\}$$

and has the properties that

$$p(x_i) \geq 0 \quad \text{and} \quad \sum_{i=1}^{\infty} p(x_i) = 1.$$

If X is continuous, it possesses a probability density function given by

$$f(x) = \frac{d}{dx} F(x).$$

The mean value of a random variable is given by

$$E(X) = \sum_{i=1}^{\infty} x_i p(x_i) \text{ in the discrete case,}$$

and

$$E(X) = \int_{-\infty}^{\infty} xf(x) \, dx \text{ in the continuous case.}$$

For the two cases, the variance is defined by

$$\text{Var}(X) = \sum_{i=1}^{\infty} [x_i - E(X)]^2 p(x_i),$$

or

$$\text{Var}(X) = \int_{-\infty}^{\infty} [x - E(X)]^2 f(x) \, dx.$$

Usually, it is more convenient to evaluate the variance from the following equivalent forms:

$$\text{Var}(X) = \sum_{i=1}^{\infty} x_i^2 p(x_i) - [E(X)]^2,$$

or

$$\text{Var}(X) = \int_{-\infty}^{\infty} x^2 f(x) \, dx - [E(X)]^2,$$

of which the square root is the standard deviation. The kth moment about the origin is denoted and defined by

$$E(X^k) = \sum_{i=1}^{\infty} x_i^k p(x_i),$$

or

$$E(X^k) = \int_{-\infty}^{\infty} x^k f(x) \, dx.$$

Note that $\text{Var}(X) = E(X^2) - [E(X)]^2$.

The random variable X has a binomial distribution if

$$P(X = k) = \binom{n}{k} p^k (1 - p)^{n-k}, \qquad p > 0 \quad \text{for } k = 0, 1, 2, \ldots, n.$$

This distribution is so-named because the underlying experiment can assume only two forms. The mean value $E(X) = np$, and the variance $\text{Var}(X) = np(1 - p)$. A second discrete distribution is the Poisson one, in which

$$P(X = k) = \frac{e^{-\lambda}\lambda^k}{k!}, \qquad k = 0, 1, 2, \ldots \quad \text{for some } \lambda > 0.$$

Both the mean and variance have value λ.

A third discrete distribution is the geometric distribution, for which

$$P(X = k) = q^{k-1}p, \qquad k = 1, 2, \ldots$$

The mean value will be found in an exercise.

The random variable X has an exponential distribution if

$$P(X \leq x) = \begin{cases} 1 - e^{-Ax}, & x \geq 0, \\ 0, & x < 0. \end{cases}$$

Here the mean $E(X) = 1/A$.

A random variable W has a normal distribution with mean μ and variance σ^2 if

$$P(W \leq w) = \frac{1}{\sqrt{2\pi\sigma^2}} \int_{-\infty}^{w} \exp\left\{ -\frac{[x - \mu]^2}{2\sigma^2} \right\} dx.$$

Note that

$$f(w) = \frac{1}{\sqrt{2\pi\sigma^2}} \exp\left\{ -\frac{[w - \mu]^2}{2\sigma^2} \right\} \quad \text{for } -\infty < w < \infty.$$

A number x_p is called a 100 pth percentile if

$$P[X < x_p] \leq p \quad \text{and} \quad P[X \leq x_p] \geq p.$$

The moment generating function of a random variable X is denoted by $M_X(\theta)$, and is given by one of the expressions:

$$M_X(\theta) = \sum_{i=1}^{\infty} e^{\theta x_i} p(x_i),$$

or

$$M_X(\theta) = \int_{-\infty}^{\infty} e^{\theta x} f(x) \, dx.$$

One use of the moment generating function appears in the formula:

$$E(X^k) = \frac{d^k}{d\theta^k} M_X(\theta)|_{\theta=0}.$$

If N is a discrete random variable, with values $0, 1, 2, \ldots$, then its probability generating function is denoted and given by

$$P(s) = \sum_{k=0}^{\infty} p_k s^k, \qquad -1 \le s \le 1.$$

The mean value $E(N) = P'(1)$. $P(s)$ also equals $E(s^N)$. Note that $P(1) = 1$.

Consider two random variables X and Y defined on the same sample space S. If for all events A and B in S, the following equation holds:

$$P(X \text{ in } A \text{ and } Y \text{ in } B) = P(X \text{ in } A)P(Y \text{ in } B),$$

we say that X and Y are independent. The same concept can be expressed through distribution functions, probability functions, or probability density functions, thus:

$$P[X \le x, Y \le y] = F_X(x)F_Y(y),$$

or

$$P[X = x, Y = y] = P[X = x]P[Y = y],$$

or

$$f(x, y) = f_X(x)f_Y(y).$$

The discrete or continuous nature of X and Y determines the appropriate equation. These equations can be extended to define the independence of n random variables $X_1, X_2, \ldots, X_n$.

For sums of independent random variables, consider

$$S_n = X_1 + X_2 + \cdots + X_n,$$

where the $\{X_i\}$ are independent random variables. Let

$$\mu_k = E(X_k), \qquad k = 1, 2, \ldots, n,$$

and

$$\sigma_k^2 = \text{Var}(X_k), \qquad k = 1, 2, \ldots, n.$$

Then

$$E(S_n) = \sum_{k=1}^{n} \mu_k \quad \text{and} \quad \text{Var}(S_n) = \sum_{k=1}^{n} \sigma_k^2.$$

Moreover the moment generating function for S_n is given by the formula

$$M_{S_n}(\theta) = M_{X_1}(\theta)M_{X_2}(\theta) \cdots M_{X_n}(\theta).$$

Likewise, the probability generating function is given by the formula

$$P_{S_n}(s) = P_{X_1}(s)P_{X_2}(s) \cdots P_{X_n}(s).$$

One form of the Central Limit Theorem will be utilized in the problems. Assume that $X_1, X_2, \ldots$ are independent, identically distributed random variables with common finite mean μ and variance σ^2. Let

$$S_n = X_1 + X_2 + \cdots + X_n.$$

Then for every fixed β,

$$\lim_{n \to \infty} P\left\{\frac{S_n - n\mu}{\sigma\sqrt{n}} \le \beta\right\} = \frac{1}{\sqrt{2\pi}}\int_{-\infty}^{\beta} e^{-x^2/2}\, dx.$$

This enables one to discuss confidence intervals for an unknown mean μ. Thus, for example:

$$P\left\{\frac{S_n}{n} - \frac{2\sigma}{\sqrt{n}} < \mu < \frac{S_n}{n} + \frac{2\sigma}{\sqrt{n}}\right\} = 0.954.$$

If σ is known then there is a probability of 0.954 that the random interval

$$\left[\frac{S_n}{n} - \frac{2\sigma}{\sqrt{n}}, \frac{S_n}{n} + \frac{2\sigma}{\sqrt{n}}\right]$$

includes the unknown mean μ. If one uses $3\sigma/\sqrt{n}$, the probability increases to 0.997.

6.2. Problems

1. Assume that the probability of a girl on a particular birth is p, and of a boy is $q = 1 - p$. Let N = random number of children for a couple with the family goal of having one girl. Find $E(N)$.

2. If $p = 0.5$ and $q = 0.5$, graphically portray the probability function for N. What is its 95th percentile?

3. Assume the same probabilities as in Problem 1. Let N = random number of children for a couple whose family goal is to have one boy and one girl. Find $E(N)$.

4. If $p = 0.5$ and $q = 0.5$, graphically portray the probability function for N in Problem 3. What is its 95th percentile?

5. Let $S = \sum_{i=1}^{5,000,000} X_i$ where each X_i represents the random number of children to a newly married couple. Find $E(S)$ under family goals from Problems 1 and 3. By how much do they differ when $p = 0.5$ and $q = 0.5$?

6. Find the moment generating function for N in Problem 1, and hence determine $\text{Var}(N)$.

7. Find the moment generating function for N in Problem 3 in general. Determine its $\text{Var}(N)$ when $p = q = 0.5$.

8. Let S be as described in Problem 5. Use the Central Limit Theorem to determine 95 percent confidence intervals for S under family strategies one and two, when $p = q = 0.5$.

9. If a couple decides to have n children—regardless of sex—what is the probability that k of these ($k \leq n$) will be girls?

10. A certain couple decides to have children until it gets one boy and will then stop having children (assuming a perfect contraceptive exists). What is the expected size of their family? What is the expected number of girls that will be born in the attempt to get one boy?

11. In a certain (hypothetical) population one third of the couples have a probability of 0.4 of bearing a boy on any birth; one third have a probability of 0.5; and one third have a probability of 0.6. If in this population all couples keep having children until they have a boy and then stop, what is the mean family size for each third of the population? What is the proportion of births that are boy children? (Goodman 1961.)

12. Refer to Problem 3.
Assuming the entire population follows such a policy determine the projected sex ratio in the population, where

$$\text{Projected sex ratio} = \frac{E(\text{no. of boys in the population})}{E(\text{no. of girls in the population})}.$$

13. (a) If the probability of a male birth equals 0.5 for the population in Question 12, what will be the expected family size in the population?
(b) How will the expected family size change if $q = 0.7$?

14. If a couple has children until it obtains two girls and the couple is able to achieve a certain degree of sex control so as to change p from 0.5 to 0.7, how will the expected family size change? Compare to Question 13.

15. In a certain population $p = 0.5$ for half the couples and $p = 0.7$ for the other half. If all couples have children until they have one child of each sex, what will be the projected sex ratio in the population? Compare to the case when $p = 0.6$ for the entire population and explain the difference. (See Questions 12 and 13.)

16. If a family has children until it obtains α sons and β daughters, what is the probability that it will have n children in total, $n \geq \alpha + \beta$.

17. The symbol p_j will represent the probability of having j sons, and assume these probabilities apply independently to all males in all generations. Let $\mu = \sum_{k=0}^{\infty} k p_k$, the expected number of direct descendants in the above branching process. Compute μ in case (a): $p_0 = 0.50$, $p_1 = 0.25$, and $p_2 = 0.25$; and case (b): $p_0 = p_1 = p_2 = p_3 = 0.25$.

18. Let x denote the chance of extinction of a male line starting with one man. Then x is the solution of Galton's equation:

$$x = p_0 + p_1 x + p_2 x^2 + \cdots$$

and relies upon the following:

"**Theorem.** *If* $\mu \leq 1$, *the process dies out with probability one. If, however,* $\mu > 1$ *the probability* x_n *that the process terminates at or before the nth generation tends to the unique root* $x < 1$ *of Galton's equation.*"

(Reference: Feller 1968, p. 297.) Utilize that theorem to determine x for the processes in Problem 17(a) and 17(b).

19. In each generation and independently a man has no sons with probability 0.1, one son with probability 0.4, two sons with probability 0.5.

(a) If we consider one male in the population who is too young to have children, what is the chance of eventual extinction of his male line?

(b) Consider a man who has two children; what is the chance of extinction of his male line?

20. In a population where one half of the men have no sons and one half have two sons, what is the probability of extinction of a male line? If one half have one son and one half have two sons, what is the chance of extinction?

21. Assume that in each generation the probabilities that a man will have i sons, $i = 0, 1, 2, \ldots$ form a geometric progression. Specifically, let the probability of a man's having no sons equal π, of having one son equal πg, of having two sons equal πg^2, etc. Determine m, the mean number of sons for this man, in terms of π and g. (Steffensen 1930.)

22. For Question 21 determine x, the probability that this man's male line will become extinct.

23. Assume a population in which the number of sons (Y) of a man follows a Poisson distribution with mean 1.1:

$$P(Y = y) = \frac{e^{-1.1}1.1^y}{y!}, \qquad y = 0, 1, 2, \ldots$$

What is the probability of extinction of a male line of descent?

24. Lotka has shown that in America the probabilities of having k sons was quite well described by the distribution $p_0 = 0.4825$, $p_k = (0.2126)$ $(0.5893)^{k-1}$ for $k \geq 1$. Note that with the exception of the first term this is a geometric progression. Using Lotka's approximations, determine the probability of extinction of a male line. (Feller 1968, p. 294.)

25. Suppose that p_j is decreased by a small amount and p_i is increased by this amount, for some $i < j$. Determine the effect on the probability of extinction.

26. Let p_i be the probability that a man will have i sons as in Question 17. Let Z_n denote the size of the nth generation (with $Z_0 = 1$), and let $f_n(s)$ be the generating function of its probability distribution. Thus, for example,

$$f_1(s) = f(s) = \sum_{i=0}^{\infty} p_i s^i.$$

Show that $f_n(s) = f(f_{n-1}(s))$.

Hint: Note that the Z_n members of the nth generation can be divided into Z_1 groups according to their ancestor in the first generation. (Feller 1968.)

27. (a) Use the result of Problem 26 to prove that $E(Z_n) = \mu^n$, where $\mu = E(Z_1)$.

(b) Compute $E(Z_n)$, $n = 1, 2, 3, 4, 5$ for the processes in Problem 17(a) and 17(b).

(c) Problem 7 (Feller 1968, pp. 301–302) yields the result

$$\text{Var}(Z_n) = \sigma^2(\mu^{2n-2} + \mu^{2n-3} + \cdots + \mu^{n-1}), \quad \text{where } \sigma^2 = \sum_{j=0}^{\infty} (j - \mu)^2 p_j.$$

Compute $\text{Var}(Z_5)$ and Standard Deviation (Z_5) for the processes in Problem 17(a) and 17(b).

28. Define a random variable

$$Y_n = 1 + Z_1 + \cdots + Z_n.$$

As explained on page 298 of Feller (1968), Y_n is "the total number of descendents up to and including the nth generation and also including the ancestor (zero generation)." As $n \to \infty$, one obtains the size of the total progeny.

(a) If $\mu < 1$, express $E(Y_n)$ in a compact form. What is $\lim_{n \to \infty} E(Y_n)$?

(b) If $\mu = 1$, express $E(Y_n)$ in a simpler form. What is $\lim_{n \to \infty} E(Y_n)$?

29. A method presented by Burks (1933) can be used to estimate the distribution of sizes of completed families in a stationary population represented by a random sample of people. Assuming that the number of families which will have n children when completed is proportional to $O(n)$, and letting $N(j)$ equal the number of children in the sample of birth order j, show under what circumstances

$$N(n) = O(n) + O(n + 1) + O(n + 2) + \cdots$$

30. (a) Using Question 29 show that $O(n) = N(n) - N(n + 1)$.

(b) Assuming that a family cannot have more than U children, determine the percentage distribution of $O(n)$, i.e., the percentage of families with n children.

31. Using $O(n)$ as described above, show that the estimated average size of completed families equals

$$N/N(1), \quad \text{where } N = \sum_{j=1}^{U} N(j).$$

$N = $ the number of children in the sample, and $N(1)$ equals the number of first born in the sample.

32. Determine the mean number of children of completed families in the population represented by this given sample:

First births	50	Fourth births	34
Second births	43	Fifth and higher	26
Third births	37	Total	190

33. Wilson and Hilferty (1935) extended the Burks rule of Question 29 by claiming that at any birth i, the number of future births to be expected can be estimated by the ratio of the number of births of order higher than i to the number of births of order i. Determine the expected number of future births for a female who has i children in the population described below and discuss the results.

Total births to Native White Mothers, by order, 1920–1933, New York State Exclusive of New York City

i	Births	i	Births	i	Births
1	339,771	8	17,112	15	468
2	248,920	9	11,157	16	233
3	155,114	10	7,312	17	120
4	95,054	11	4,405	18	61
5	59,445	12	2,697	19	34
6	38,641	13	1,618	20+	39
7	25,610	14	926		

34. Assume that a random sample of adults in a population is obtained and each person is asked how many siblings he or she has. Letting N_k equal the number of families with k children, estimate P_k, the proportion of all families—with at least one child—which contain k children.

35. Assume that a random sample of 28 adults produced $N_0 = 2$, $N_1 = 6$, $N_2 = 12$, and $N_3 = 8$. Use the proportions $\{P_k\}$ to compute the mean number of children in a completed family in the population.

36. The probability R_k that a family with k children will increase (i.e., the probability that a couple will go on to their $(k + 1)$th child after having had their kth child) is called the parity progression ratio. Show that

$$ R_k = \frac{\sum_{i=k+1}^{\infty} P_i}{\sum_{i=k}^{\infty} P_i} $$

and define P_k in terms of R_k.

37. For the sample described in Problem 35 estimate the parity progression ratios.

38. If $R_0 = a$ is the chance of having a first child for married couples, and $R_i = b$ is the chance of having an $(i + 1)$th child for couples with i children, $i > 0$, what is the distribution of children among married couples? What is the mean number of children to married couples? (Ryder.)

39. Consider a time-homogeneous Markov chain with states of nature u (urban dweller) and r (rural dweller). Let the matrix of transition probabilities be:

$$M = \begin{pmatrix} m_{uu} & m_{ur} \\ m_{ru} & m_{rr} \end{pmatrix} = \begin{pmatrix} 0.999 & 0.001 \\ 0.020 & 0.980 \end{pmatrix}.$$

Assume that the initial distribution of urban and rural dwellers is:

$$P^{(0)} = \begin{pmatrix} P_u \\ P_r \end{pmatrix}.$$

The superscript (0) on P denotes the initial time. Future times will also be indicated by superscripts. In order to project forward, interchange rows and columns,

$$\begin{pmatrix} m_{uu} & m_{ru} \\ m_{ur} & m_{rr} \end{pmatrix} \begin{pmatrix} P_u \\ P_r \end{pmatrix} = \begin{pmatrix} m_{uu}P_u + m_{ru}P_r \\ m_{ur}P_u + m_{rr}P_r \end{pmatrix}.$$

In general, the distribution t years later is given by

$$P^{(t)} = \tilde{M}^t P^{(0)},$$

where $\tilde{M}$ is the transpose of M. If $P_u = 8{,}000{,}000$ and $P_r = 77{,}000{,}000$, find $P^{(10)}$.

Reference: This problem, as well as Problems 40 through 46, are based on the paper: Beekman, John A. "Several demographic projection techniques," *Rural Demography*, **8** (1981), 1–11.

40. The Markovian model in Problem 39 has no provision for growth. For simplicity, assume that the growth factor $e^r = e^{0.02}$ can be applied in each of the 10 years. The population projections then become:

$$P^{(1)} = \tilde{M}e^r P^{(0)},$$

$$P^{(2)} = \tilde{M}e^r P^{(1)} = \tilde{M}e^r \tilde{M}e^r P^{(0)} = \tilde{M}^2 e^{2r} P^{(0)},$$

$$P^{(3)} = \tilde{M}e^r P^{(2)} = \tilde{M}e^r \tilde{M}^2 e^{2r} P^{(0)} = \tilde{M}^3 e^{3r} P^{(0)}, \ldots,$$

$$P^{(10)} = \tilde{M}^{10} e^{10r} P^{(0)}.$$

Compute $P^{(10)}$ for Problem 39 with $r = 0.02$.

41. The Markovian model in Problems 39 and 40 can show the effects of rural–urban shifts on aggregate births. If the overall birth rates in urban and rural parts are b_u and b_r, respectively, arranged in a horizontal vector B, then the births at time t are

$$BP^{(t)} = (b_u \quad b_r)\begin{pmatrix} m_{uu} & m_{ru} \\ m_{ur} & m_{rr} \end{pmatrix}^t \begin{pmatrix} e^{tr}P_u \\ e^{tr}P_r \end{pmatrix}.$$

Assume that $b_u = 0.035$ and $b_r = 0.050$. Compute $Be^{10r}P^{(0)} - B\tilde{M}^{10}e^{10r}P^{(0)}$ to see the difference in aggregate projected births for the eleventh year from the initial time point.

42. The statistical estimation of the transition probabilities of a Markov chain can be based on the principle of maximum likelihood (see, for example, Ross 1972, pp. 240–242). Such an estimator of the probability that the process goes from state i to state j is obtained from the proportion of time that the process, when leaving state i, next enters state j. How would you estimate m_{ur} and m_{ru}? Use a property of M to deduce estimates for m_{uu} and m_{rr}.

43. Powers of the M matrix (rather than the transpose) give probabilities of events happening within various lengths of time. Interpret the elements of M^{10}.

44. Let

$$\pi_u = \lim_{n \to \infty} m_{iu}^{(n)} \quad \text{for } i = u \text{ or } r;$$

$$\pi_r = \lim_{n \to \infty} m_{ir}^{(n)} \quad \text{for } i = u \text{ or } r.$$

Our Markov chain is ergodic (see page 95 of the S. Ross book), and hence π_u and π_r exist independent of the value of i, and they constitute the unique nonnegative solution vector of the equations:

$$\pi_u = \pi_u m_{uu} + \pi_r m_{ru},$$

$$\pi_r = \pi_u m_{ur} + \pi_r m_{rr},$$

$$\pi_u + \pi_r = 1.$$

Solve that system for π_u and π_r. How would you interpret π_u and π_r?

45. For the matrix M in Problem 39, obtain π_u and π_r. Assume a different set of matrix values, i.e.,

$$M^* = \begin{pmatrix} 0.999 & 0.001 \\ 0.004 & 0.996 \end{pmatrix}.$$

What revised values of π_u and π_r result?

46. The ergodic theorem for Markov chains (Problem 44 above) has a parallel in Lotka's ergodic theorem in demography. What does it say, and why are the theorems similar?

47. Assume that $\{X_n, n = 0, 1, 2, \ldots\}$ is a Markov chain. Let the n-step probabilities be

$$P_{ij}^n = P\{X_{n+m} = j \mid X_m = i\}$$

for $n \geq 0$, $i, j \geq 0$. The Chapman–Kolmogorov equations state that

$$P_{ij}^{n+m} = \sum_{k=0}^{\infty} P_{ik}^{n} P_{kj}^{m}$$

for all $n, m \geq 0$, all i, j.

(a) Prove those equations.

(b) Graphically illustrate what they say.

Reference: This problem, as well as Problems 48 and 49, are based on Chapters 4 and 6 of the S. Ross text.

48. A birth and death process is a system whose state at any time is the number of people in the system at that time. It can be shown that when there are n people in the system then the time until the next arrival is exponentially distributed with mean $1/\lambda_n$, and is independent of the time until the next departure which is itself exponentially distributed with mean $1/\mu_n$. Prove that if one event occurs at an exponential rate λ and another independent event at an exponential rate μ, then together they occur at an exponential rate $\lambda + \mu$.

49. Let $X(t)$ be the number of people in a birth and death process at time t, and let P_n be the limiting probability that there will be n people in the system, i.e.,

$$P_n = \lim_{t \to \infty} P\{X(t) = n\}.$$

On pages 154–155 of the S. Ross text, it is demonstrated that

$$P_n = \frac{\lambda_0 \lambda_1 \cdots \lambda_{n-1}}{\mu_1 \mu_2 \cdots \mu_n \left(1 + \sum_{n=1}^{\infty} \frac{\lambda_0 \lambda_1 \cdots \lambda_{n-1}}{\mu_1 \mu_2 \cdots \mu_n} \right)}$$

for $n \geq 1$. For these probabilities to exist, it is necessary and sufficient that

$$\sum_{n=1}^{\infty} \frac{\lambda_0 \lambda_1 \cdots \lambda_{n-1}}{\mu_1 \mu_2 \cdots \mu_n} < \infty.$$

Assume that

$$\mu_n = n\mu, \qquad n \geq 1,$$
$$\lambda_n = n\lambda + \theta, \qquad n \geq 0,$$

where $\lambda > 0$, $\mu > 0$, and $\theta > 0$. Each person is assumed to give birth at an exponential rate λ. Also, there is an exponential rate of immigration of θ. Hence, by Problem 48, the total birth rate when there are n persons in the system is $n\lambda + \theta$. What conditions are needed for the test series to converge?

50. Assume that $\lambda_n = n\lambda$, $\mu_n = n\mu$ for $n \geq 0$, $n \geq 1$, respectively, and that

$$P_n(t) = P\{X(t) = n\}.$$

Then $P_n(t)$ satisfies the differential equations

$$P_0'(t) = \mu P_1(t),$$

$$P_n'(t) = -(\lambda + \mu)nP_n(t) + \lambda(n-1)P_{n-1}(t) + \mu(n+1)P_{n+1}(t).$$

Define

$$M(t) = \sum_{n=1}^{\infty} nP_n(t).$$

Assume that the initial population size is A. Prove that $M(t) = Ae^{(\lambda - \mu)t}$. What conclusion do you reach, as $t \to \infty$?
Reference: Pages 456–457 of W. Feller text.

51. Following (abridged from Lotka) is a distribution of completed fraternities, i.e., a distribution of fathers according to the number of their sons.

Size of fraternity	Number of cases
0	5
1	2
2	1
3	1
4	1
Total	10

Find
  (a) the mean number of sons for a father taken at random, say G;
  (b) the variance σ^2 of the number of sons;
  (c) the distribution of sons according to the number of brothers that they have;
  (d) the mean number of brothers for a son taken at random, say B.

52. (a) Show that the relation

$$B = \frac{\sigma^2}{G} + G - 1$$

holds for your results in the preceding question.
  (b) Show that it holds in general.
  (c) Given that $B = 2$ and $G = 1$, calculate the standard deviation of completed fraternities.

53. The official estimate of population x at some future time for a particular town is (thousands)

x	Probability
250	1/3
275	1/2
300	1/6

The losses $L(x, \hat{x})$ associated with combinations of x and $\hat{x}$ are known to be

		$\hat{x}$		
		250	275	300
	250	0	9	21
x	275	30	0	16
	300	72	30	0

Find what estimate $\hat{x}$ gives minimum loss, as among 250, 275, 300.

6.3. Solutions

1.

$$
\begin{aligned}
E(N) &= 1p + 2qp + 3q^2p + \cdots \\
&= (1 - q) + 2q(1 - q) + 3q^2(1 - q) + \cdots \\
&= 1 + q + q^2 + q^3 + \cdots \\
&= 1/(1 - q) = 1/p.
\end{aligned}
$$

2.

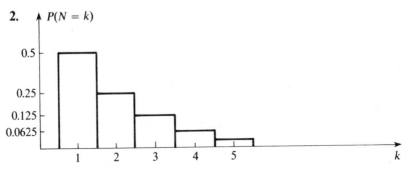

The 95th percentile is five.

3.

$$
\begin{aligned}
E(N) &= 2(pq + qp) + 3(p^2q + q^2p) + 4(p^3q + q^3p) + 5(p^4q + q^4p) + \cdots \\
&= q\{2p + 3p^2 + 4p^3 + 5p^4 + \cdots\} + p\{2q + 3q^2 + 4q^3 + 5q^4 + \cdots\} \\
&= q\frac{d}{dp}\sum_{k=2}^{\infty} p^k + p\frac{d}{dq}\sum_{k=2}^{\infty} q^k \\
&= q\frac{d}{dp}\left\{\frac{1}{1 - p} - 1 - p\right\} + p\frac{d}{dq}\left\{\frac{1}{1 - q} - 1 - q\right\} \\
&= q^{-1} - q + p^{-1} - p = q^{-1} + p^{-1} - 1.
\end{aligned}
$$

Move briefly, if N_b is the number of children until the first boy, and N_g the number until the first girl, and N_{bg} the number until one boy or one girl, then

$$E(N) = E(N_b) + E(N_g) - E(N_{bg})$$

$$= \frac{1}{q} + \frac{1}{p} - 1.$$

4. The 95th percentile is six.

5. $E(S) = 5,000,000/p$, and $E(S) = 5,000,000(q^{-1} + p^{-1} - 1)$. When $p = 0.5$, and $q = 0.5$, the mean amounts differ by 5,000,000!

6.

$$E[e^{\theta N}] = \sum_{k=1}^{\infty} e^{\theta k} q^{k-1} p$$

$$= \frac{pe^{\theta}}{1 - qe^{\theta}}, \qquad \theta < -\ln_e q.$$

$$\mathrm{Var}(N) = \frac{q}{p^2}.$$

7.

$$E[e^{\theta N}] = \sum_{k=2}^{\infty} e^{\theta k} \{qp^{k-1} + pq^{k-1}\}$$

$$= \frac{q}{p} \sum_{k=2}^{\infty} (pe^{\theta})^k + \frac{p}{q} \sum_{k=2}^{\infty} (qe^{\theta})^k$$

$$= \frac{qpe^{2\theta}}{1 - pe^{\theta}} + \frac{pqe^{2\theta}}{1 - qe^{\theta}} \text{ for } \theta < \min(-\ln_e p, \, -\ln_e q).$$

When $p = q = 0.5$, $E[e^{\theta N}] = e^{2\theta}/(2 - e^{\theta})$, and

$$\mathrm{Var}(N) = E(N^2) - [E(N)]^2$$
$$= 11 - 9 = 2.$$

8. For family strategy number one,

$$E(S) \pm 2 \text{ Standard Deviation}(S) = 10,000,000 \pm 2(10,000,000)^{1/2}$$
$$= 10,000,000 \pm 6,325.$$

For family strategy number two,

$$E(S) \pm 2 \text{ Standard Deviation}(S) = 15,000,000 \pm 2(10,000,000)^{1/2}$$
$$= 15,000,000 \pm 6,325.$$

9. $P[X = k] = \binom{n}{k} p^k (1 - p)^{n-k}.$

10. (Dual of Problem 1.) $E(N) = 1/q$. $E(N - 1) = q^{-1} - 1$.

12. Let there be C couples in the population. Then

$$\text{Projected sex ratio} = \frac{C(q^{-1} + p^{-1} - 1)q}{C(q^{-1} + p^{-1} - 1)p} = \frac{q}{p}.$$

13. (a) 3.0.   (b) 3.76.

14. The expected number of children until the first girl is $1/p$; then the process starts over with another $1/p$ children to the second girl; hence a total of $2/p$ children.

If $p = 0.5$, $E(N) = 4$, whereas if $p = 0.7$, $E(N) = 2.86$.

15. As in Problem 12, let there be C couples in the population. In the first case,

$$\text{Projected sex ratio} = \frac{\frac{1}{2}C(3)(0.5) + \frac{1}{2}C(3.76)(0.3)}{\frac{1}{2}C(3)(0.5) + \frac{1}{2}C(3.76)(0.7)} = 0.636.$$

In the second case,

$$\text{Projected sex ratio} = \frac{C(3.1667)(0.4)}{C(3.1667)(0.6)} = 0.667.$$

The average family size in the first case is 3.38, and a slightly greater percentage will be girls than in the second case.

16.

$$P[N = n] = P[N = n, \text{ and the last child born } = \text{girl}]$$

$$+ P[N = n, \text{ and the last child born } = \text{boy}]$$

$$= \frac{(n - 1)!}{(\beta - 1)!\,(n - \beta)!} p^{\beta - 1} q^{n - \beta} p + \frac{(n - 1)!}{(\alpha - 1)!\,(n - \alpha)!} q^{\alpha - 1} p^{n - \alpha} q.$$

17. (a) 0.75.   (b) 1.5.

18. (a) $x = 1$.   (b) $-1 + \sqrt{2} \doteq 0.4142$.

19. (a) $x = 0.2$ $(\mu = 1.4)$.

(b) It is 1 if he (already) has two girls; it is 0.2 if he has one son; and it is $0.2(0.2) = 0.04$ if he has two sons.

20. (a) Since $\mu = 1$, $x = 1$.

(b) $\mu = 1.5$, and $x = 0$. From first principles, the chance of extinction is 0.

21.

$$m = \sum_{i=0}^{\infty} i\pi g^i = \sum_{i=1}^{\infty} i\pi g^i = \pi g \frac{d}{dg} \sum_{i=1}^{\infty} g^i = \pi g(1 - g)^{-2}.$$

22.

$$x = [1 - (1 - 4g\pi)^{1/2}]/(2g),$$

provided $\pi > (1 - g)^2/g$, $x = 1$ otherwise.

23.

$$x = \sum_{y=0}^{\infty} \frac{e^{-1.1}(1.1)^y}{y!} x^y = e^{-1.1}e^{1.1x} = e^{-1.1(1-x)}.$$

Since $\mu = 1.1$, the desired $x < 1$. As a first trial value, let $x_1 = 0.9$. Successive values are found by the Newton–Raphson method (see Kellison (1975)), with

$$f(x) = x - e^{-1.1(1-x)}.$$

The results are 0.614, 0.757, 0.810, 0.8230.

24.

$$x = p_0 + \sum_{k=1}^{\infty} 0.2126(0.5893)^{k-1}x^k.$$

After summing the series, the resulting equation in x is

$$0.5893\, x^2 - 1.0717\, x + 0.4825 = 0.$$

The meaningful root is 0.82.

25. Let $p_k^* = p_k$, all k, except $p_j^* = p_j - \varepsilon$, and $p_i^* = p_i + \varepsilon$. Then

$$x = \sum_{k=0}^{\infty} p_k^* x^k = \sum_{k=0}^{\infty} p_k x^k + \varepsilon(x^i - x^j).$$

Since $x^i - x^j > 0$, the new probability of extinction is greater. This also follows from general reasoning.

26. Z_n is the sum of Z_1 random variables $Z_n^{(k)}$, each representing the size of the offspring of one member of the first generation. By a conditioning argument, the generating function for Z_n is

$$f_n(s) = E\{s^{Z_n}\} = \sum_{j=0}^{\infty} E\{s^{Z_n}|Z_1 = j\}P\{Z_1 = j\}.$$

By assumption the $Z_n^{(k)}$ are independent, and each has the same probability distribution as Z_{n-1}. Hence,

$$E\{s^{Z_n}|Z_1 = j\} = \{E\{s^{Z_{n-1}}\}\}^j.$$

Hence

$$f_n(s) = \sum_{j=0}^{\infty} \{E\{s^{Z_{n-1}}\}\}^j P\{Z_1 = j\}$$

$$= \sum_{j=0}^{\infty} \{f_{n-1}(s)\}^j p_j$$

$$= f(f_{n-1}(s)).$$

27. (a)

$$E(Z_n) = \frac{d}{ds} f_n(s)\big|_{s=1}$$

$$= \frac{d}{ds} f(f_{n-1}(s))\big|_{s=1}$$

$$= f'(f_{n-1}(1))f'_{n-1}(1)$$

$$= f'(1)f'_{n-1}(1)$$

$$= \mu E(Z_{n-1}).$$

Note that $E(Z_0) = 1$. Hence $E(Z_n) = \mu^n$.

(b) For the process in Problem 17(a), the expected values are 0.75, 0.5625, 0.4219, 0.3164, and 0.2373.

For the process in Problem 17(b), the expected values are 1.5, 2.25, 3.375, 5.0625, and 7.5938.

(c) For process Problem 17(a),

$$\sigma^2 = 0.6875, \qquad \text{Var}(Z_5) = 0.6636 \quad \text{and} \quad \text{Standard Deviation}(Z_5)$$

$$= 0.8146.$$

For process Problem 17(b),

$$\sigma^2 = 1.250, \qquad \text{Var}(Z_5) = 83.4521 \quad \text{and} \quad \text{Standard Deviation}(Z_5)$$

$$= 9.1352.$$

28. (a)

$$E(Y_n) = 1 + \mu + \cdots + \mu^n = \frac{1 - \mu^{n+1}}{1 - \mu}.$$

$$\lim_{n \to \infty} E(Y_n) = \frac{1}{1 - \mu}.$$

(b)

$$E(Y_n) = n + 1. \qquad \lim_{n \to \infty} E(Y_n) = +\infty.$$

29. For better understanding, write out the equations for $N(1)$, $N(2)$, and $N(3)$: e.g., $N(2)$ = number of children of birth order $2 = O(2) + O(3) + O(4) + \cdots$ Such relations are true provided that a complete enumeration is made of each sampled family. Thus, a sample point is one family's children, not a randomly selected child.

30. (a) Express $N(n)$ and $N(n + 1)$ in terms of the $O(k)$'s and subtract both sides of the equations.

(b) We want $A \sum_{n=1}^{U} O(n) = 100$. But $A \sum_{n=1}^{U} O(n) = AN(1)$. Hence, $A = 100/N(1)$, and $100\, O(n)/N(1)$ is the percentage of families with n children, $n = 1, 2, \ldots, U$.

31.

$$\text{Average size of families} = \sum_{k=1}^{U} kp_k$$

$$= \sum_{k=1}^{U} k \frac{O(k)}{N(1)}$$

$$= \frac{1}{N(1)} \sum_{k=1}^{U} k[N(k) - N(k+1)]$$

$$= \frac{1}{N(1)} [1N(1) - 1N(2) + 2N(2) - 2N(3)$$

$$+ 3N(3) - 3N(4) + \cdots]$$

$$= \frac{N}{N(1)}.$$

32. $N/N(1) = 3.8$. Note that $O(1) = 7$, $O(2) = 6$, $O(3) = 3$, $O(4) = 8$, and $O(5) = 26$.

33. Let $r_i = \sum_{j>i} j$ Births/(i Births)

i	r_i	i	r_i	i	r_i
1	1.969	8	1.699	15	1.041
2	1.687	9	1.606	16	1.090
3	1.708	10	1.450	17	1.117
4	1.787	11	1.407	18	1.197
5	1.858	12	1.297	19	1.147
6	1.858	13	1.163	20+	0.000
7	1.803	14	1.031		

These numbers do not seem to be realistic estimators of future births.

34. $P_k = \alpha(N_k/k)$. Thus for the data of the following Problem 35 $P_1 = \alpha\frac{2}{1}$, $P_2 = \alpha\frac{6}{2}$, $P_3 = \alpha\frac{12}{3}$, and $P_4 = \alpha\frac{8}{4}$. Since $\sum_{i=1}^{4} P_i = 1$, $\alpha = 11$.

35.

$$E(N) = 1(\tfrac{2}{11}) + 2(\tfrac{3}{11}) + 3(\tfrac{4}{11}) + 4(\tfrac{2}{11}) = 2.55.$$

36. $R_k = \sum_{i=k+1}^{\infty} P_i / \sum_{i=k}^{\infty} P_i$ by the definition of conditional probability.

$P_U = R_{U-1}$;

$R_1 = (1 - P_1)/1$ and hence $P_1 = 1 - R_1$;

$R_2 = (1 - P_1 - P_2)/(1 - P_1)$ and hence $P_2 = R_1(1 - R_2)$;

$R_3 = (1 - P_1 - P_2 - P_3)/(1 - P_1 - P_2)$ and hence $P_3 = R_1 R_2(1 - R_3)$.

In general, $P_k = R_1 R_2 \cdots R_{k-1}(1 - R_k)$.

37. $R_1 = \frac{9}{11}; R_2 = \frac{2}{3}; R_3 = \frac{1}{3}; R_4 = 0.$

38.

$$R_0 = a \Rightarrow P_0 = 1 - a; P_1 = (1 - b)a; P_2 = b(1 - b)a;$$
$$P_3 = b^2(1 - b)a; \ldots; P_k = b^{k-1}(1 - b)a.$$

$$E(N) = \sum_{k=1}^{\infty} kb^{k-1}(1 - b)a = (1 - b)a \frac{d}{db}\left(\sum_{k=0}^{\infty} b^k\right) = \frac{a}{1 - b}.$$

39.

$$\tilde{M}^2 = \begin{pmatrix} 0.99802 & 0.03958 \\ 0.00198 & 0.96042 \end{pmatrix},$$

and

$$\tilde{M}^{10} = \begin{pmatrix} 0.99089 & 0.18212 \\ 0.00911 & 0.81788 \end{pmatrix}.$$

Thus

$$P^{(10)} = \tilde{M}^{10}P^{(0)} = \begin{pmatrix} 21,950 \\ 63,050 \end{pmatrix}.$$

40.

$$P^{(10)} = \tilde{M}^{10}e^{10r}P^{(0)} = \tilde{M}^{10}\begin{pmatrix} 9,771 \\ 94,048 \end{pmatrix} = \begin{pmatrix} 26,810 \\ 77,009 \end{pmatrix}.$$

41.

$$Be^{10r}P^{(0)} = (0.035 \quad 0.050)\begin{pmatrix} 9,771 \\ 94,048 \end{pmatrix} = 5,044$$

$$B\tilde{M}^{10}e^{10r}P^{(0)} = (0.035 \quad 0.050)\begin{pmatrix} 26,810 \\ 77,009 \end{pmatrix} = 4,789.$$

The difference is 255.

42.

$$\hat{m}_{ur} = \frac{\text{Twelve Month Rural Increase} - \text{Natural Rural Increase}}{\text{July 1 Estimate of Urban Population}},$$

$$\hat{m}_{ru} = \frac{\text{Twelve Month Urban Increase} - \text{Natural Urban Increase}}{\text{July 1 Estimate of Rural Population}}.$$

$$\hat{m}_{uu} = 1 - \hat{m}_{ur} \quad \text{and} \quad \hat{m}_{rr} = 1 - \hat{m}_{ru}.$$

43.

$m_{uu}^{(10)}$ = the probability of starting as an urban dweller, and after 10 periods being an urban dweller.

$m_{ur}^{(10)}$ = the probability of starting as an urban dweller, and after 10 periods being a rural dweller.

$m_{ru}^{(10)}$ = the probability of starting as a rural dweller, and after 10 periods being an urban dweller.

$m_{rr}^{(10)}$ = the probability of starting as a rural dweller, and after 10 periods being a rural dweller.

44.

$$\pi_u = m_{ru}/(m_{ur} + m_{ru}),$$
$$\pi_r = m_{ur}/(m_{ur} + m_{ru}).$$

π_u and π_r are the long-term proportions of urban and rural people in a population.

45.

$$\pi_u = 0.020/[0.001 + 0.020] = 0.9524,$$
$$\pi_r = 0.001/[0.001 + 0.020] = 0.0476.$$

For M^*, $\pi_u = 0.800$, and $\pi_r = 0.200$.

46. Lotka's theorem says that:

"*A population subjected to fixed rates of mortality and fertility will asymptotically approach an age distribution which depends only on the schedule of mortality and fertility.*"

The theorems are similar because the long-term distributions (probability or age) are independent of the initial states of nature or age distributions.

47. (a)

$$P_{ij}^{n+m} = P\{X_{n+m} = j \mid X_0 = i\}$$

$$= \sum_{k=0}^{\infty} P\{X_{n+m} = j, X_n = k \mid X_0 = i\}$$

by the law of total probability,

$$= \sum_{k=0}^{\infty} P\{X_{n+m} = j \mid X_n = k, X_0 = i\} P\{X_n = k \mid X_0 = i\}$$

by a conditioning argument,

$$= \sum_{k=0}^{\infty} P_{ik}^n P_{kj}^m \quad \text{by the Markovian property.}$$

(b)

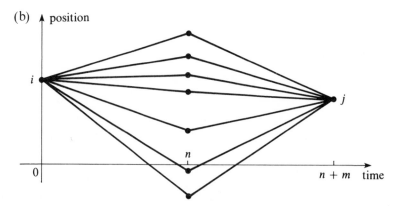

48. Let T_1 = time at which first event occurs and T_2 = time at which second event occurs. Our assumptions are that

$$P\{T_1 \le t\} = 1 - e^{-\lambda t},$$
$$P\{T_2 \le t\} = 1 - e^{-\mu t}.$$

Let T = time when either T_1 or T_2 occurs, i.e., $T = \min(T_1, T_2)$. Then

$$
\begin{aligned}
P\{T \le t\} &= 1 - P\{T > t\} \\
&= 1 - P\{\min(T_1, T_2 > t)\} \\
&= 1 - P\{T_1 > t, T_2 > t\} \\
&= 1 - P\{T_1 > t\}P\{T_2 > t\} \text{ by independence,} \\
&= 1 - e^{-\lambda t}e^{-\mu t} \\
&= 1 - e^{-(\lambda + \mu)t}.
\end{aligned}
$$

Hence, T has an exponential distribution with rate $\lambda + \mu$.

49. For this process,

$$\sum_{n=1}^{\infty} \frac{\lambda_0 \lambda_1 \cdots \lambda_{n-1}}{\mu_1 \mu_2 \cdots \mu_n} = \sum_{n=1}^{\infty} \frac{\theta(\theta + \lambda) \cdots (\theta + (n-1)\lambda)}{\mu^n n!}.$$

By the ratio test for series, this will converge when

$$\lim_{n \to \infty} \frac{\theta(\theta + \lambda) \cdots (\theta + n\lambda)}{(n+1)! \mu^{n+1}} \cdot \frac{n! \mu^n}{\theta(\theta + \lambda) \cdots (\theta + (n-1)\lambda)}$$

$$= \lim_{n \to \infty} \frac{\theta + n\lambda}{(n+1)\mu} = \frac{\lambda}{\mu} < 1.$$

This is true when $\lambda < \mu$. When $\lambda \ge \mu$, one can demonstrate that the necessary and sufficient condition for convergence is not fulfilled.

50. Multiply the second differential equation by n, and add over $n = 1, 2, \ldots$. The terms involving n^2 cancel, and one obtains

$$M'(t) = \lambda \sum_{n=1}^{\infty} (n-1)P_{n-1}(t) - \mu \sum_{n=0}^{\infty} (n+1)P_{n+1}(t) = (\lambda - \mu)M(t).$$

A little algebra is needed to justify the second series starting at $n = 0$. The differential equation for $M(t)$, with initial condition

$$P_A(0) = 1 \quad (\text{or } M(0) = A)$$

yields

$$M(t) = Ae^{(\lambda - \mu)t}.$$

This function tends to 0, A, or ∞, according as $\lambda < \mu$, $\lambda = \mu$, or $\lambda > \mu$.

51. (a) $G = 1.1$.
(b) $\sigma^2 = E(N^2) - (1.1)^2 = 1.89$.
(c)

Sons with k brothers	Number of cases
0	2
1	2
2	3
3	4
Total	11

(d) $B = 1.818$.

53. The losses are:

$$L_1 = \tfrac{1}{3}(0) + \tfrac{1}{2}(30) + \tfrac{1}{6}(72) = 27,$$
$$L_2 = \tfrac{1}{3}(9) + \tfrac{1}{2}(0) + \tfrac{1}{6}(30) = 8,$$

and

$$L_3 = \tfrac{1}{3}(21) + \tfrac{1}{2}(16) + \tfrac{1}{6}(0) = 15.$$

Therefore, estimate $\hat{x} = 275$ has minimum loss.

Brief List of References

Works on Demography

Cox, Peter R. (1976), *Demography* (5th Ed.), Cambridge University Press: Cambridge.
Keyfitz, Nathan (1968), *Introduction to the Mathematics of Population*, Addison-Wesley: Reading, Mass.*
Keyfitz, Nathan and Wilhelm Flieger (1968), *World Population: An Analysis of Vital Data*, University of Chicago Press: Chicago.
Keyfitz, Nathan (1977), *Applied Mathematical Demography*, Wiley: New York.
Pollard, A. H., Farhat Yusuf, and G. N. Pollard (1981), *Demographic Techniques* (2nd Ed.), Pergammon Press: Sydney.
Pressat, Roland (1972), *Demographic Analysis*, Aldine–Atherton: Chicago (English translation of *L'Analyse Demographique*, Presses Universitaires de France: Paris, 1969).
Shryock, Henry S. and Jacob S. Siegel (1971), *The Methods and Materials of Demography*, Vols. 1 and 2, U.S. Government Printing Office: Washington, D.C.
Spiegelman, M. (1968), *Introduction to Demography* (Revised Ed.), Harvard University Press: Cambridge, Mass.

A small number of problems have brief references of the form (Burks 1933). Complete bibliographic details will be found in the references sections of the above books.

Actuarial Texts

Batten, R. W. (1978), *Mortality Table Construction*, Prentice-Hall: Englewood Cliffs, New Jersey.
Bowers, Newton L., Hans U. Gerber, James C. Hickman, Donald A. Jones, and Cecil J. Nesbitt (1983), *Actuarial Mathematics*, Society of Actuaries: Chicago.

---

* A revised printing of this reference was made in 1977. Those revisions included a new Chapter 20, "Problems in the Mathematics of Population," pages 425–444. Many of those problems have been included in this book, with the permission of Addison-Wesley Publishing Co. It should be noted that those pages contained few solutions, whereas this book has at least as many pages devoted to solutions as to problems.

Greville, T. N. E. (1974), *Part 5 Study Notes on Graduation, Society of Actuaries:* Chicago.

Jordan, C. W. (1967), *Life Contingencies* (2nd Ed.), Society of Actuaries: Chicago.

Miller, M. D. (1946), *Elements of Graduation*, Society of Actuaries: Chicago.

Further References

Beekman, John A. (1974), *Two Stochastic Processes*, Almqvist and Wiksell: Stockholm, also Halsted Press (Wiley): New York.

Elandt-Johnson, Regina C. and Norman L. Johnson (1980), *Survival Models and Data Analysis*, Wiley: New York.

Feller, William (1968), *An Introduction to Probability Theory and Its Applications*, Vol. I (3rd Ed.), Wiley: New York.

Halmos, P. R. (1980), The heart of mathematics, *The American Mathematical Monthly*, **87**, 519–524.

Kellison, Stephen G. (1975), *Fundamentals of Numerical Analysis*, R. D. Irwin, Inc: Homewood, IL.

Ralston, Anthony (1965), *A First Course in Numerical Analysis*, McGraw-Hill: New York.

Ross, Sheldon M. (1972), *Introduction to Probability Models*, Academic Press: New York.

Scarborough, J. B. (1958), *Numerical Mathematical Analysis*, Johns Hopkins Press: Baltimore.

Index

Problem Books in Mathematics *(continued)*

A Problem Seminar
by *Donald J. Newman*

Exercises in Number Theory
by *D.P. Parent*